AF755039

Philipp Wolfrum

Information Routing, Correspondence Finding, and Object Recognition in the Brain

Studies in Computational Intelligence, Volume 316

Editor-in-Chief
Prof. Janusz Kacprzyk
Systems Research Institute
Polish Academy of Sciences
ul. Newelska 6
01-447 Warsaw
Poland
E-mail: kacprzyk@ibspan.waw.pl

Further volumes of this series can be found on our homepage: springer.com

Vol. 294. Manuel F.M. Barros, Jorge M.C. Guilherme, and Nuno C.G. Horta
Analog Circuits and Systems Optimization based on Evolutionary Computation Techniques, 2010
ISBN 978-3-642-12345-0

Vol. 295. Roger Lee (Ed.)
Software Engineering, Artificial Intelligence, Networking and Parallel/Distributed Computing, 2010
ISBN 978-3-642-13264-3

Vol. 296. Roger Lee (Ed.)
Software Engineering Research, Management and Applications, 2010
ISBN 978-3-642-13272-8

Vol. 297. Tania Tronco (Ed.)
New Network Architectures, 2010
ISBN 978-3-642-13246-9

Vol. 298. Adam Wierzbicki
Trust and Fairness in Open, Distributed Systems, 2010
ISBN 978-3-642-13450-0

Vol. 299. Vassil Sgurev, Mincho Hadjiski, and Janusz Kacprzyk (Eds.)
Intelligent Systems: From Theory to Practice, 2010
ISBN 978-3-642-13427-2

Vol. 300. Baoding Liu (Ed.)
Uncertainty Theory, 2010
ISBN 978-3-642-13958-1

Vol. 301. Giuliano Armano, Marco de Gemmis, Giovanni Semeraro, and Eloisa Vargiu (Eds.)
Intelligent Information Access, 2010
ISBN 978-3-642-13999-4

Vol. 302. Bijaya Ketan Panigrahi, Ajith Abraham, and Swagatam Das (Eds.)
Computational Intelligence in Power Engineering, 2010
ISBN 978-3-642-14012-9

Vol. 303. Joachim Diederich, Cengiz Gunay, and James M. Hogan
Recruitment Learning, 2010
ISBN 978-3-642-14027-3

Vol. 304. Anthony Finn and Lakhmi C. Jain (Eds.)
Innovations in Defence Support Systems, 2010
ISBN 978-3-642-14083-9

Vol. 305. Stefania Montani and Lakhmi C. Jain (Eds.)
Successful Case-Based Reasoning Applications-1, 2010
ISBN 978-3-642-14077-8

Vol. 306. Tru Hoang Cao
Conceptual Graphs and Fuzzy Logic, 2010
ISBN 978-3-642-14086-0

Vol. 307. Anupam Shukla, Ritu Tiwari, and Rahul Kala
Towards Hybrid and Adaptive Computing, 2010
ISBN 978-3-642-14343-4

Vol. 308. Roger Nkambou, Jacqueline Bourdeau, and Riichiro Mizoguchi (Eds.)
Advances in Intelligent Tutoring Systems, 2010
ISBN 978-3-642-14362-5

Vol. 309. Isabelle Bichindaritz, Lakhmi C. Jain, Sachin Vaidya, and Ashlesha Jain (Eds.)
Computational Intelligence in Healthcare 4, 2010
ISBN 978-3-642-14463-9

Vol. 310. Dipti Srinivasan and Lakhmi C. Jain (Eds.)
Innovations in Multi-Agent Systems and Applications – 1, 2010
ISBN 978-3-642-14434-9

Vol. 311. Juan D. Velásquez and Lakhmi C. Jain (Eds.)
Advanced Techniques in Web Intelligence, 2010
ISBN 978-3-642-14460-8

Vol. 312. Patricia Melin, Janusz Kacprzyk, and Witold Pedrycz (Eds.)
Soft Computing for Recognition based on Biometrics, 2010
ISBN 978-3-642-15110-1

Vol. 313. Imre J. Rudas, János Fodor, and Janusz Kacprzyk (Eds.)
Computational Intelligence in Engineering, 2010
ISBN 978-3-642-15219-1

Vol. 314. Lorenzo Magnani, Walter Carnielli, and Claudio Pizzi (Eds.)
Model-Based Reasoning in Science and Technology, 2010
ISBN 978-3-642-15222-1

Vol. 315. Mohammad Essaaidi, Michele Malgeri, and Costin Badica (Eds.)
Intelligent Distributed Computing IV, 2010
ISBN 978-3-642-15210-8

Vol. 316. Philipp Wolfrum
Information Routing, Correspondence Finding, and Object Recognition in the Brain, 2010
ISBN 978-3-642-15253-5

Philipp Wolfrum

Information Routing, Correspondence Finding, and Object Recognition in the Brain

Springer

Dr. Philipp Wolfrum
Siemens AG
Corporate Technology
Otto-Hahn-Ring 6
81739 Munich
Germany
E-mail: wolfrum@fias.uni-frankfurt.de

ISBN 978-3-642-15253-5 e-ISBN 978-3-642-15254-2

DOI 10.1007/978-3-642-15254-2

Studies in Computational Intelligence ISSN 1860-949X

Library of Congress Control Number: 2010934961

© 2010 Springer-Verlag Berlin Heidelberg

This work is subject to copyright. All rights are reserved, whether the whole or part of the material is concerned, specifically the rights of translation, reprinting, reuse of illustrations, recitation, broadcasting, reproduction on microfilm or in any other way, and storage in data banks. Duplication of this publication or parts thereof is permitted only under the provisions of the German Copyright Law of September 9, 1965, in its current version, and permission for use must always be obtained from Springer. Violations are liable to prosecution under the German Copyright Law.

The use of general descriptive names, registered names, trademarks, etc. in this publication does not imply, even in the absence of a specific statement, that such names are exempt from the relevant protective laws and regulations and therefore free for general use.

Typeset & Cover Design: Scientific Publishing Services Pvt. Ltd., Chennai, India.

Printed on acid-free paper

9 8 7 6 5 4 3 2 1

springer.com

Acknowledgements

This book would not have been possible without the people I have had the honor of working with over the past years. Major parts of the book originate from my doctoral studies at the Frankfurt Institute for Advanced Studies. My first thanks therefore go to my academic advisor, Christoph von der Malsburg. His scientific ideas have inspired my research and had a strong influence on the concepts developed in this book. Discussions with Geoff Goodhill, Steve Grossberg, Bruno Olshausen, Rudolf Mester, Jochen Triesch, and Alan Yuille, among others, have shaped and widened my thinking. I'm sure you will find the traces of our discussions in here! I also thank all colleagues at FIAS for good company and the truly interdisciplinary interaction we had. Special thanks go to the members of our group, Urs Bergmann, Jenia Jitsev, Jörg Lücke, and Junmei Zhu. Last but not least I owe thanks to my parents, who planted curiosity and a desire for understanding in me that ultimately lead to this book.

Foreword

At first sight, this book is about face recognition in the brain. Its more lasting value, however, lies in the paradigmatic way in which this particular problem is treated. From the basic ideas that are worked out here in concrete detail, it is a natural and simple next step to at least imagine, if not realize in model form, much more general structures and processes, thus helping to bridge the still tremendous chasm between mind and brain. It is the purpose of this foreword to point out these generic traits.

For centuries, thinking about the brain has been dominated by the most complex mechanistic devices of the time, clockwork, communicating hydraulic tubes or, today, the computer. The computer, taken as incarnation of the Universal Turing Machine, can implement any conceivable process, so that also a functional brain can surely be simulated on it, an idea that, beginning in the fifties of the last century, has been seducing scientists to create "artificial intelligence" in the computer. As a result we now have an information technology that displays many functional capabilities formerly regarded as the exclusive domain of the mind. As fascinating as this is, doting on "intelligent machines" is systematically diverting our attention away from the true problems of understanding the working of the brain.

"Intelligence in the machine" is, of course, in truth canned human intelligence. The computer itself is without a clue, behaving like a blind man guided by a distant voice step by step over a tight-rope. It is humans that are exploiting their own insight into complex processes to endow the computer with algorithms, reaction patterns that cover those and only those situations the programmer has thought of ahead of time. From the technological point of view this would be fine, if the complexity of it all weren't beginning to grow out of hand. Little noticed by the general public, for a decade or two there has been growing awareness among the experts of a "software crisis" that is bugging information technology. Large software projects have an alarmingly high likelihood of complete failure, and if not failure, of time and cost overruns and an abundance of functional deficits. The venture of artificial intelligence has completely underestimated the complexity of such processes as language or motion control or vision, and it is evident by now that no reasonable effort can generate such capabilities directly in algorithmic fashion. Realizing this, instead of running the computer with borrowed human intelligence we should rather concentrate on the organization processes that create functional structures in the child's brain (or indeed in the programmer's mind), implement them on our computers and let them do the rest of the work.

The basis for the programmer's ability to create algorithms is creative infrastructure comprising world knowledge, sample structures, methodology as well as the ability to project goals, diagnose mistakes or interpret the symbols handled in the machine. It is this infrastructure that enables humans to formulate abstract goals and to translate these abstract goals into concrete process descriptions. Understanding this infrastructure will enable us to understand mind and brain, and it will be the basis for another profound technological revolution.

As misleading as the computer is when using human intelligence as a shortcut to problem solving, as instrumental can it be as a tool to model the processes going on in the brain. Once we understand creative mind processes, the computer, being Turing-universal, will be able to emulate them. Information technology may then retract from the details in the machine, to concentrate fully on the formulation of our human goals and intentions. This will not only liberate humans from programming chores, setting free their creativity in terms of goal formulation, but will also liberate the computing medium from its leash and let it deal with challenges as they arise, meeting them by reference to the abstract goals we set instead of to distant human foresight.

Trying to model mind processes in the computer is a powerful tool to study the fundamental issues of the neural and cognitive sciences. It is a false expectation that experiments can force functional ideas on us. Experimentation needs guidance by hypotheses and beliefs. Now, it is a sad observation that many highly convincing functional ideas, some of them the inspiration of extensive experimental paradigms, turn out to be flawed when tried on the computer. It therefore is a wise strategy to develop functional computer models with generic brain features, and to demonstrate their functional viability "experimentally". They will powerfully influence thinking about the brain and, given time, will inspire neuroscientific experiments, just as Rosenblatt's perceptron once inspired Hubel and Wiesel's experimental paradigm.

This book, essentially the PhD thesis of its author, is focused on one such function, the invariant representation and recognition of visual structure. It brings together a number of threads that have been developed over decades in my lab. The model that is developed in these pages offers conceptual solutions to a number of fundamental issues regarding brain function. Among these are, first and foremost, the four cognitive architecture questions — what are the data structures of the brain on the fast time scale of brain state and on the slow time scale of memory; and, what are the mechanisms of organization, again on both time scales, brain state and memory?

To this day, the neurosciences are dominated by a certain set of answers to these questions, beginning with a very specific idea regarding the data structure of brain state. According to this, neurons are taken as elementary symbols (blue light in position x, an edge of orientation ϕ in position y, muscle fiber z to be twitched etc.), and the brain's state in a given moment is thought to be completely described by the vector of neural activities. Further, brain states are organized by exchange of excitation and inhibition between neurons, regulated such that they evolve slowly from quasi-stationary state to quasi-stationary state, each lasting for a tenth of a fifth of a second. The data structure of memory has the form of strengths of neural connections (synaptic weights), and the mechanism of organization of memory is constituted by different forms of synaptic plasticity.

Modelling of brain functions, a very active field in recent decades, has almost exclusively been based on this conceptual framework, and although a great wealth of results has sprung forth from this effort, one cannot help to suspect that progress is held back by some conceptual roadblock. In the mid-eighties the enthusiasm of the field of artificial neural networks reached the sky, claiming that in a few years' time information technology would be revolutionized and the brain understood, but although activity levels in neural theory and such related fields as machine learning, Bayesian estimation or robotics are higher than ever today, those great aspirations have not materialized, no one daring to come forth with predictions when they will. In some quarters, one might even sense a feeling of depression and rising doubt back-engineering of the brain could be possible in a few decades, given that evolution had taken so long to generate it.

There is, however, reason to be optimistic. Cerebral cortex, our cognitive organ, is estimated to contain 5×10^{10} neurons, and each neuron to have 10^4 synaptic connections. It would take about 10^{16} bits (or one PetaByte) to describe a list of these connections. In contrast, there are only about 10^9 bits of information in the human genome (only a small part of which can be assumed to code the brain's wiring diagram). Also a lifetime of learning is not enough to close the gap. Generously counting 30 years of growing up, which amounts to 3×10^8 seconds of waking time, and generously assuming the brain to absorb 300 bits of sensory information each second, only 10^{11} bits of information could come from that source. This leaves us short by a factor of 10^5 of being able to account for the information content of the list. The conclusion of this little calculation is that to an overwhelming extent the connectivity patterns of the brain are dominated by regularity. (Random connectivity is no alternative, in spite of the neuroanatomist's impression under the microscope, and in spite of some of it being invoked by certain theorists, as coordination between different brain structures would be severely hampered by ubiquitous independent random decisions.) Part of this regularity is due to spatial organization, neurons connecting mainly within a relatively small neighborhood. If each synapse has only to choose from 10^n neurons in its environment (with n likely to be between 4 and 5) instead of all neurons in cortex, the information requirement is reduced from 10^{16} bits to $n \times 10^{15}$ bits, still leaving a tremendous gap. The tremendous remaining amount of structural regularity is the hunting ground of brain theory.

The most promising approach to closing the information gap is trying to understand the process of development of the nervous system. Studying this is a very active field, and has been for over a hundred years. In distinction to our human technology, where the manufacture of artefacts starts with a blueprint of the whole, and parts are shaped and assembled by outside forces, life constructs its organisms inside out, the movers being the component molecules and cells, the final Gestalt emerging gradually in the process in a progression from coarse to fine. This development is all driven by the behavioral repertoire of individual cells, inherited, refined and differentiated in a continuous line from our single-celled ancestors. Cells divide and multiply, differentiate, shape themselves, move, adhere selectively to each other, put out extensions and communicate with each other by sending and receiving molecular, mechanical and electrical signals in ever-richer patterns. Thus, growth of the embryo, of the brain, of the detailed connectivity of the nervous system, is a complex dance of cells, choreographed from moment to moment by the molecular network of genetic regulation. One way of looking at it is to see the organism as a party joyfully celebrated by its population of cells.

In distinction to human artefacts, which are totally passive during design and assembly and start to perform only after being finished, the body, nervous system and molecular signalling web of the organism are functioning from the outset, and this function is essential for shaping development and even design. Development progresses from simple to complex in a process of gradual differentiation, with more and more detail emerging in stages. The enormous number of degrees of freedom that are generated in this process are dealt with as they arise, system dynamics being regulated such that it is of contracting nature, an aspect that Waddington called channeling, such that most variables tend towards set points, annihilating initial and any accidental deviations. Only a very small set of variables is unstable, with a tendency to grow, leaving never more than a few choices, and it is at these branching points that the genes exert their essential influence. In a breath-taking development over the last two decades molecular biology has discovered a set of genes that to a large extent is common to all animals, from worm to human. The experts speak of the ontogenetic toolkit, the piano on which evolution lighthandedly plays animal tunes.

When it comes to the nervous system, after development of the neural tube and its various bulbar swellings, cells differentiate into neurons and they begin to grow processes with which to connect up with each other, starting the game of network self-organization that will continue into adulthood. Initially, cells reach out with their processes into their neighborhood, but then they send them out to more distant targets, guided by molecular signals that themselves are created by other cells. These signals form complex patterns that are shaped by the growing network and in turn are shaping the network. As result of this feedback, networks converge towards patterns that are stabilized by the very signals they put out themselves. Over the last three decades, some prime examples for this process have been studied intensively, prominent among them the establishment of neighborhood-preserving, retinotopic, fiber projections between eye and brain.

Network self-organization, probably the most pervasive and fundamental process by which Life gives itself shape from molecular to societal levels, is in some respects akin to crystallization. A crystal is shaped by the preferences that individual molecules have concerning their immediate neighborhood. Just as these local preferences conspire in a crystal to create long-range order, local preferred network arrangements conspire to form globally ordered networks. Local preferences may differ in detail, depending on molecular signals and genetic influences, but dominant among them are likely to be a tendency towards sparsity (relatively few incoming and outgoing connections per neuron) and cooperativity between alternate pathways (such as the pathways A→C and A→B→C, or the pathways A→B→C and A→B'→C). During network self-organization, spatial neighborhood relationships, like within retina or within cortex, will play a potent influence, evident in the relative abundance of short-range connections. However, a neuron's neighbors in the network sense are all those other neurons to whom it is connected — and these may be at any distance in the spatial sense. Many aspects and details of self-organized networks therefore may not be evident to anatomists looking through the microscope, the reason why they speak of random connectivity. They may be random in the spatial sense, but they aren't in the sense of connectivity.

It is a very interesting and important question how the attractor states of network self-organization, the connectivity patterns that dominate the structure of the nervous system, may be characterized in terms of their global structure. There may be no general answer to this question, or at least not any time soon. However, it is possible to characterize an important class of attractor networks. Imagine a two- or three-dimensional space populated by neurons with short range connections in this space. Such a network will produce signal ensembles that are dominated by short-range correlations, which in turn will favor short-range connections. It is therefore natural to expect that such "topological networks" are attractor states. In spite of its simple construction, the universe of such networks is complex and interesting, as many different topological networks can be created on the same total set of neurons, each individual network having as support a different subset of all neurons, supporting a combinatorially rich family of networks.

Network self-organization takes place on many temporal scales, the evolutionary growth of gene-regulatory networks being the slowest (although it may happen under the biologist's eye when studying bacterial evolution). Much faster are the processes of network self-organization during ontogenesis and learning. A taste of this can be found in Chapter 5 of this book, where the ontogenesis of the so-called switchyard circuits is described as a process of network self-organization.

At the heart of this book, however, is the hypothesis that network self-organization takes place also on the psychological time scale of about a tenth of a second, and that rapidly switching connections form an integral part of the data structure of brain state, in importance on a par with neural activity switching, the classical neural data structure. According to this view, the permanent, physical connections formed by slow synaptic plasticity during ontogenesis or learning can be switched off or on again, in a matter of milliseconds. The activity state of a connection means two things, on the one hand, an active connection is conducting and transmits signals, on the other it expresses the statement that the two connected neurons are related to each other, that they form a composite symbol.

A symbol system that wants to be expressive has to exploit combinatorics, building up more complex symbols from more elementary ones. In order to be able to do so, the system has to have some kind of glue, by which elements can be attached to each other. All known symbol systems have this capability. Thus, by arranging elements spatially on paper we form characters out of strokes, words out of characters etc., or mathematical expressions out of elementary symbols. Neurons in the brain cannot move about to form composite symbols, and there correspondingly has to be another means of building up higher structures. This, the hypothesis goes, is done with the help of switching synapses.

In this book, two applications of this idea play an important role. To build up the representation of a piece of visual structure, textural elements in their two-dimensional spatial arrangement are attached to each other by activated links. Having this means of expression is particularly important in memory, where novel representations are to be fluently composed out of pre-existing structural pieces under the guidance by retinal input. The other application is the center-piece of a recognition mechanism based on point-to-point correspondences between incoming images and stored object models, as modelled in detail in Chapter 3.

If switching connections are an important aspect of the data structure of brain state there must be a mechanism to organize and control these variables. This can be achieved by a rapid, reversible kind of Hebbian plasticity: the synapse between two neurons with signals that are correlated on a fine time scale is activated (or stabilized in its active resting state), whereas the synapse is switched off temporarily if the two neurons are both active but uncorrelated in their activity. On this basis, a process of network self-organization can take place, by which, for instance, the set of links that establish the correspondence between image and model can be activated or the representation of a complex scene can be glued together in a structured way.

The *de novo* organization of a complex network with many connections is a complicated affair, and if it was to be based entirely on signal correlation-controlled synaptic plasticity it would be relatively slow, for two reasons. On the one hand, neural signals have low bandwidth and the temporal resolution of signal coincidences is surely not better than one millisecond. When many connections are to be sorted into the active or passive state, it takes a large number of time slices of that width to make that distinction for them all. On the other hand, as the signal correlations themselves are to be shaped by the network, they are confused during the early stages of network organization and a longish process of gradual differentiation is required to reach the organized state. Thus, organized network structure is a valuable resource which, once formed, should be stored for quick later retrieval.

Such storage is indeed possible by finding and storing associations between sets of synapses that during signal correlation-controlled network self-organization are switched on simultaneously. This novel process, modelled in (Bergmann and von der Malsburg n.d.), is analogous to the formation of neural assemblies by an associative mechanism of Hebbian type, and one might speak of synaptic assemblies. To the extent that the synapses to be associated are contacting different target neurons a specific communication medium between synapses is required. This medium is probably constituted by a particular class of cells, which would correspond to the control units proposed by Charles Anderson (Olshausen et al. 1993), modelled as associative mechanism in (Lücke and von der Malsburg 2006) and further developed and used intensively in Chapter 3 of this book. Several possibilities for physiological realization are discussed in Chapter 4. An idea to be considered is that control units are to be identified with astrocytes (Möller et al. 2007).

The ideas expressed in this foreword may or may not be convincing to you, feeling that intuitively appealing arguments are too easily offered, and armchair modelling goes wrong too often. It therefore is important to take pains and work out models in all concreteness — even if experiment will later prove many or most of the details wrong — so that the functional viability of the underlying ideas can be tested on the computer. This is the significance of the work presented here. For the sake of concreteness it had to concentrate on some specific application, face recognition, and had to limit itself to mere sketches of the processes going on in the brain or thought to be going on in the brain. The lasting value of this work is to be found, however, in it generic traits. The most fundamental among these is the proposal that dynamic links, rapidly switching synapses, are an integral part of the data structure of brain state. Given that for each cortical neuron there are an estimated 10,000 synapses, the classical view, which restricts attention to one activity variable per neuron, would be ignoring 99.99% of all the information needed to describe that state. What this book is advocating is a profound paradigm shift, opening the door to vistas not even imaginable within the classical view.

One of these is the ability of active networks to represent the structure of mental objects. The reality that we experience is in our head, of course, and is a construction of our mind, although a construction guided by sensory input. This reality of the present scene is to a very large extent ephemeral, no two scenes ever being identical in detail. The representation of our inner reality is composed by the brain as a collage of structural pieces collected in the past. Dynamic links are the glue with which to put pieces together in a structured way.

Sets of active neurons, the classical data structure, have no inherent structure, structure that would allow, for instance, to judge whether two such sets stand in any kind of meaningful relationship (other than having been associated with each other in the past, indicated by associatively strengthened mutual permanent connections). By contrast, two well-organized dynamic graphs may or may not be composable to a larger well-organized dynamic graph again. In the example given in this book, a stored model network may or may not be homomorphic to an active network in primary visual cortex. Both networks have two-dimensional topological structure, and they are homomorphic if there is a topological mapping between them connecting cells labeled with similar features. This correspondence can be discovered by the system the first time the two dynamic nets are activated. The basis for this is previous training of control structures that connect small sub-networks (small sets of neighboring feature-specific neurons) here and there, sub-networks that are re-used again and again as part of different larger networks. Whereas in this book links are allowed to be dynamic only between stored model and image, when flexible models are to be constructed in short-term memory as a collage of stored sub-networks, dynamic links will be required to glue them together, and structural relationships must be defined to decide whether two subnetworks fit together or not, even if they are to be combined for the first time.

This book intends to demonstrate the viability of a set of qualitative ideas as model for the brain. Important further steps are necessary to complete the paradigm shift. One is neuroscientific validation of the basic aspects of the theory, among them the rapid reversible switching of synapses under signal control, the existence and structure of control units, and the generic nature of self-organized dynamic networks. A second, corresponding to T. Kuhn's puzzle solving, is the massive amount of work required to interpret the elements of cognitive science in terms of the machinery advocated here. And finally, it will be necessary to develop the mathematical apparatus that alone can turn what in retrospect will one day look like tinkering into a coherent theory of the working of brain and mind.

Frankfurt, April 2010 Christoph von der Malsburg

References

Bergmann, U., von der Malsburg, C.: Self-organization of topographic bilinear networks for invariant recognition. Neural Computation (Under revision)

Lücke, J., von der Malsburg, C.: Rapid correspondence finding in networks of cortical columns. In: Kollias, S.D., Stafylopatis, A., Duch, W., Oja, E. (eds.) ICANN 2006. LNCS, vol. 4131, pp. 668–677. Springer, Heidelberg (2006)

Möller, C., Lücke, J., Zhu, J., Faustmann, P.M., von der Malsburg, C.: Glial cells for information routing? Cognitive Systems Research 8(1), 28–35 (2007)

Olshausen, B.A., Anderson, C.H., van Essen, D.C.: A neurobiological model of visual attention and invariant pattern recognition based on dynamic routing of information. Journal of Neuroscience 13(11), 4700–4719 (1993)

Contents

List of Figures

Chapter 1
Introduction

A central task of the brain is finding homomorphies or correspondences between patterns. When we look at a scene, for example, our visual system has to find correspondence between the pattern that falls onto the retina and memories stored in the brain, enabling us to make sense of our environment. This ought to work even when the image we get to see of an object in a certain situation differs significantly from its representation in memory, which was created under very different conditions.

Such differences in appearance can arise from projective effects, environmental influences, and intrinsic changes to the object itself. Projective effects occur due to a changed position of the retina (i.e. the projection surface) with respect to the unchanged object, resulting in translation, scaling, 2D or 3D rotation of the image on the retina. Environmental influences such as a changed light source result in different reflections, shadows, or lighting conditions. Finally, the object itself can change, for example its shape or the expression of a person's face.

All these instances result in changes of the image projected onto the retina. For recognition, this new retinal pattern must be matched with known memory patterns residing in the brain, and correspondences with the correct memory must be established. Thus, the process of visual correspondence finding cannot simply compare the two patterns in a direct way, but the comparison needs to be invariant to differences that do not change the *meaning* of those patterns.

Finding correspondences invariantly of transformations is not only important in vision. It is also required for sensory tasks, e.g. perceiving speech and music: You want to understand words independently of the pitch and pace of a speaker, you want to recognize a melody regardless of the musical instrument it is played on. The important role of correspondence finding becomes most obvious in high level tasks like abstract reasoning: If we have understood a certain grammatical concept, we can easily apply it to new vocabulary or even another language that follows the same rules. For this, we need to find correspondences between the schema representing the grammatical concept and the concrete sentence we are currently confronted with.

When *acting* instead of perceiving, the brain needs to invert this process. It may start out, e.g., with the plan of grasping an object in front of us, and then has to translate this high-level plan into the corresponding, very specific motor commands

P. Wolfrum: Info. Routing, Corres. Finding, and Object Recognition in the Brain, SCI 316, pp. 1–3.
springerlink.com © Springer-Verlag Berlin Heidelberg 2010

to the different arm and hand muscles. These motor patterns will turn out to be quite different when the object we want to grasp has a different weight or surface structure, and they will look different again if we want to use a tool instead of the hand alone. Using its understanding of the correspondences between two situations, the brain has to generate from the known motor pattern of one situation the correct one for the new situation.

As we can see, correspondence finding is crucial for all kinds of information processing in the brain, and it may be a key factor to understanding intelligence per se. There exist very good textbooks that cover our current understanding of the brain in a general way, and I don't think I could contribute very much by adding another one. I do believe, however, that a book that starts out from the *functional concept* of correspondence finding can open up new ways of looking at how the brain works that might otherwise go unnoticed. This belief serves me as justification for writing this book.

To exemplify the concept of correspondence finding, I will mostly focus on visual information processing in the brain. The reasons for this are twofold. First, the visual areas are the most well understood parts of the mammalian brain. Important general principles as well as huge quantities of details are known about visual information processing in the brain, much more than for any other sensory modality. This large amount of information has resulted in an equally large number of computational models addressing visual processing in the brain. Second, the independent and large scientific field of *computer vision* (CV) also addresses the question of visual information processing, but (usually) from a purely functional, computational perspective. As we will see later, the approaches taken in CV may turn out as useful hypotheses for neural function, while concepts learnt from the brain can stimulate CV in turn. We therefore devote a large part of Chapter 2 to an introduction to the primate visual system and to different neural and computational models of vision. The remainder of Chapter 2 is devoted to a discussion of these models, which will deepen our understanding of the concept of correspondence finding and set the stage for the other parts of the book.

As we will see in Chapter 2, correspondence finding poses two main requirements to the brain as a physical and computational system:

- Signal pathways must exist in the brain that allow routing of information between different parts of the brain, such that the patterns residing there can be compared. These pathways have to be manifold enough to allow the many types of routing and transformations we hinted at above; on the other hand they should be as parsimonious as possible for evolutionary reasons.
- The brain must have computational mechanisms that implement the actual correspondence finding using these pathways. Realistically, this process has to be structured into several stages, to achieve high flexibility without drowning in a combinatorial explosion. These stages have to cooperate since the task of one alone can usually not be carried out without solving the full problem.

These requirements give rise to a multitude of questions: How can correspondence finding be implemented in the brain? What kinds of information routing pathways are advantageous? How can such structures self-organize during ontogenesis? And how can all this be integrated into a functional object recognition system?

After having set the stage in Chapter 2, we will turn to some of these questions in detail in the subsequent chapters. In Chapter 3, we develop a neurally plausible model for object recognition. In that chapter, we make very simple assumptions on the underlying routing structures and rather concentrate on the correspondence finding mechanisms. Chapter 4, on the other hand, focuses on the connectivity infrastructure necessary for correspondence finding. It argues for the need of multistage routing structures and introduces an architecture that is optimal in terms of required neural resources. In Chapter 5 we argue that complex neural structures must arise from rather simple developmental rules, and we derive a mechanism that can explain the ontogenesis of the previously introduced routing structures. Chapter 6 outlines how the ideas developed in the previous chapters can be combined into an integrated system performing routing over several stages *and* object recognition. We conclude the book with a discussion and outlook in Chapter 7.

Chapter 2
Background and Concepts

2.1 The Primate Visual System

When we look at the object in front of us, a specific pattern of activity is created in the ganglion cells of the retina. This pattern is relayed and transformed on its way via the thalamus and primary visual areas to higher cortical stages, where it may interact with and activate certain memories stored there. If this happens, we feel that we have *recognized* the object. While the recognition process as a whole is far from being understood, there is a wealth of details known about the individual anatomical subsystems involved in this process. Light entering the eye from the environment is focussed and projected by the lens as an inverted image onto the back of the eye. This concave surface is covered by the retina, the first outpost of the central nervous system (CNS) to be encountered by the light (see Figure 2.1a). Here, the light first traverses several cellular layers before it hits the photoreceptors and is transformed into neuronal signals. There are two classes of photoreceptors: *Rods* can function under very poor lighting conditions; a single photon can elicit a measurable response in a rod. Several rods propagate their signals to the same subsequent cell, thus further increasing amplification while at the same time reducing spatial resolution of the resulting signal. Rods are responsible for the gray, low-resolution images we perceive at night. *Cones*, on the other hand, have a poorer sensitivity, allowing them only to operate under day light-like conditions. However, they are very densely packed in the fovea (the spot of the retina providing highest resolution), and there are three types of cones sensitive to different wavelengths of light. Thus, the cones produce the sharp and colorful images we see during the day in the center of our view. The photoreceptors pass on their continuous voltage signals to several interneurons (bipolar, horizontal, and amacrine cells), from where they are propagated to the retinal ganglion cells. Resulting from the intricate wiring from photoreceptors to ganglion cells, some ganglion cells respond to bright spots on dark background ("on-center-off-surround"), while others respond to dark centers surrounded by light ("off-center-on-surround"). Some cells (called M cells) have large receptive fields (RFs) and high temporal sensitivity, while others (P cells) have small, precisely located RFs with low temporal sensitivity (see Figure 2.1b).

P. Wolfrum: Info. Routing, Corres. Finding, and Object Recognition in the Brain, SCI 316, pp. 5–27.
springerlink.com
© Springer-Verlag Berlin Heidelberg 2010

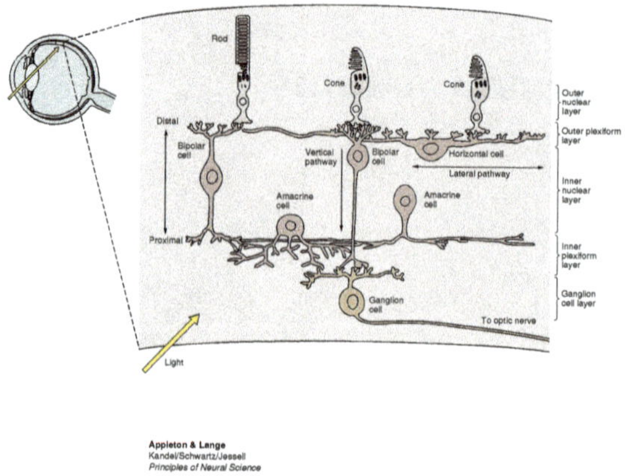

(a)

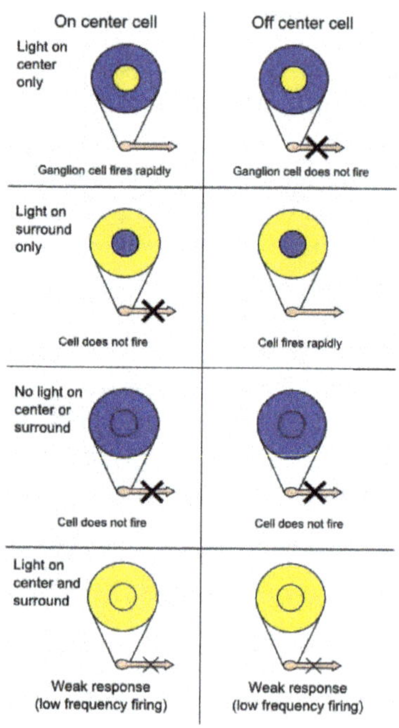

(b)

Fig. 2.1 Visual processing in the retina. **(a)** After traversing the eye and the retina, light is detected by the photoreceptors sitting at the back of the retina. The photoreceptors pass their signals via several interneurons to the ganglion cells. Reprinted from (Kandel et al., 2000) with the permission of McGraw-Hill. **(b)** The specific connectivity of the retinal cells leads to two kinds of ganglion receptive fields. On-center cells respond most strongly when only the central part of their RF receives light, while off-center cells respond most to stimulation of the surround of their RF.

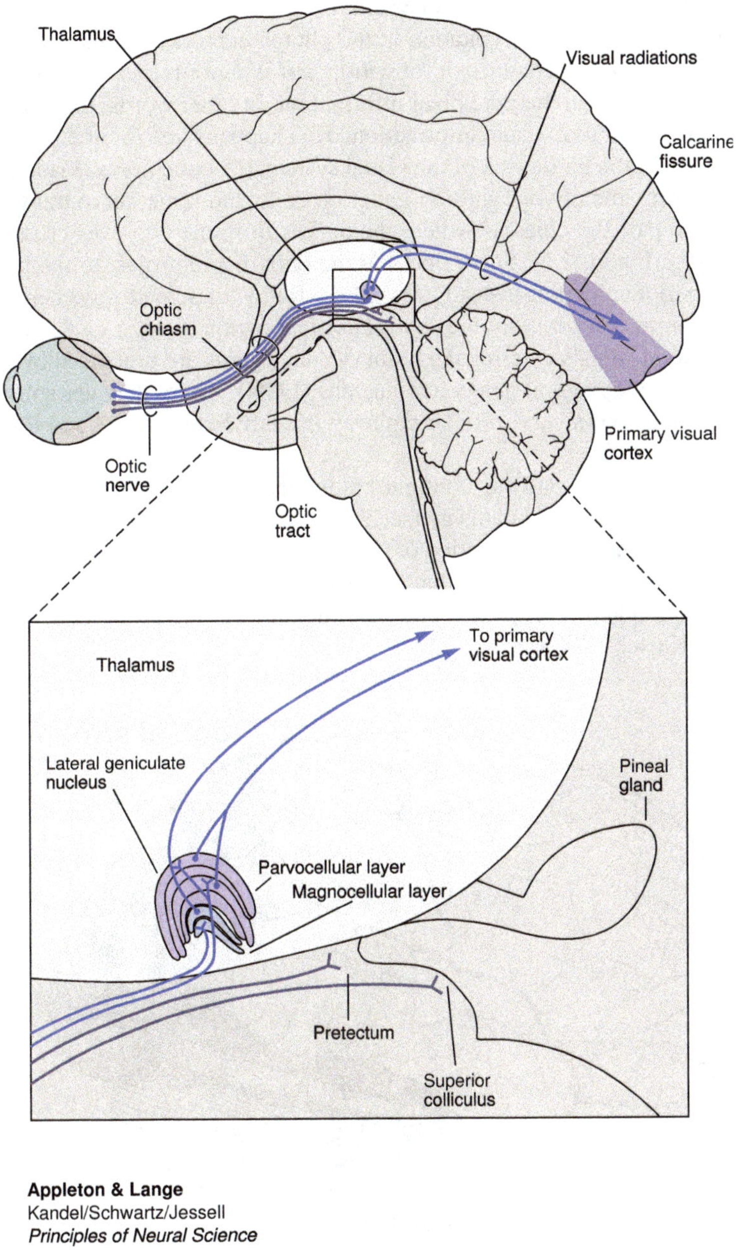

Fig. 2.2 The way from the retina to cortex. The optic nerve relays the output of the ganglion cells to the thalamus, from where the signals travel to the primary visual cortex. Reprinted from (Kandel et al., 2000) with the permission of McGraw-Hill.

Thus, already here the coding of visual information has changed from a computer screen-like pixel representation in the photoreceptors to signals representing contrasts and temporal changes, information that is more relevant for understanding a visual scene than the pure pixel information. In other words, there has been a lossless compression of visual information. This happens just in time because now the signals enter the bottleneck of the visual system, the optic nerve. Made up of the axons of approximately one million ganglion cells, the optic nerve transports the visual image past the chiasm—where the nerves from the two eyes cross—to the thalamus (see Figure 2.2). Some fibers of the optic nerve project to the pretectum and the superior colliculus instead; we will not follow these pathways here.

The thalamus is a centrally located subcortical region acting as a "gateway" for most peripheral signals entering the brain. Visual signals are processed by a part of the thalamus called lateral geniculate nucleus (LGN), where they are split up into a magnocellular and a parvocellular pathway (mainly fed by the M and P ganglion cells, respectively).

The magnocellular pathway continues along the upper part of the cortex and is therefore also called dorsal pathway (see Figure 2.3). While the dorsal pathway primarily analyzes motion and position of objects in a scene, we will focus here on the parvocellular (or ventral) stream that is concerned mostly with object recognition. Note, however, that this separation of tasks is not quite as unique in reality (see, e.g., Konen and Kastner, 2008).

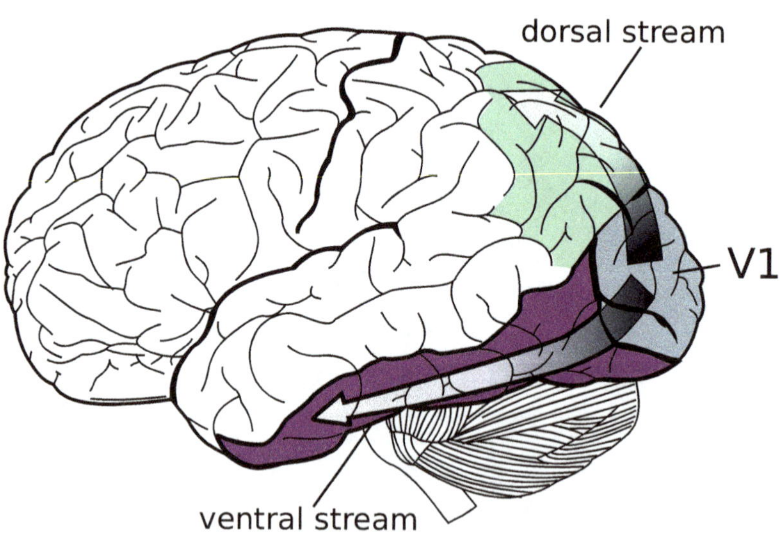

Fig. 2.3 Ventral and dorsal stream. The ventral stream (shown in dark) extends along the lower part of the cortex from the occipital towards the temporal areas. The dorsal stream (light gray) also starts in the occipital cortex, but continues dorsally into the parietal lobe.

The ventral stream extends along the lower part of the cortex from the occipital towards the temporal areas. From LGN, it first reaches the primary visual cortex V1 (also called *striate cortex*; here and in the following, we use the naming of cortical areas in the Macaque brain, which do not correspond perfectly to the human brain but are much better understood than their human counterparts.). V1 is located in the very rear of the occipital cortex around the calcarine sulcus (cf. Figure 2.3). A V1 simple cell receives input from several center-surround LGN cells arranged in a row and thus becomes sensitive to *edges* of contrast at a certain position and orientation

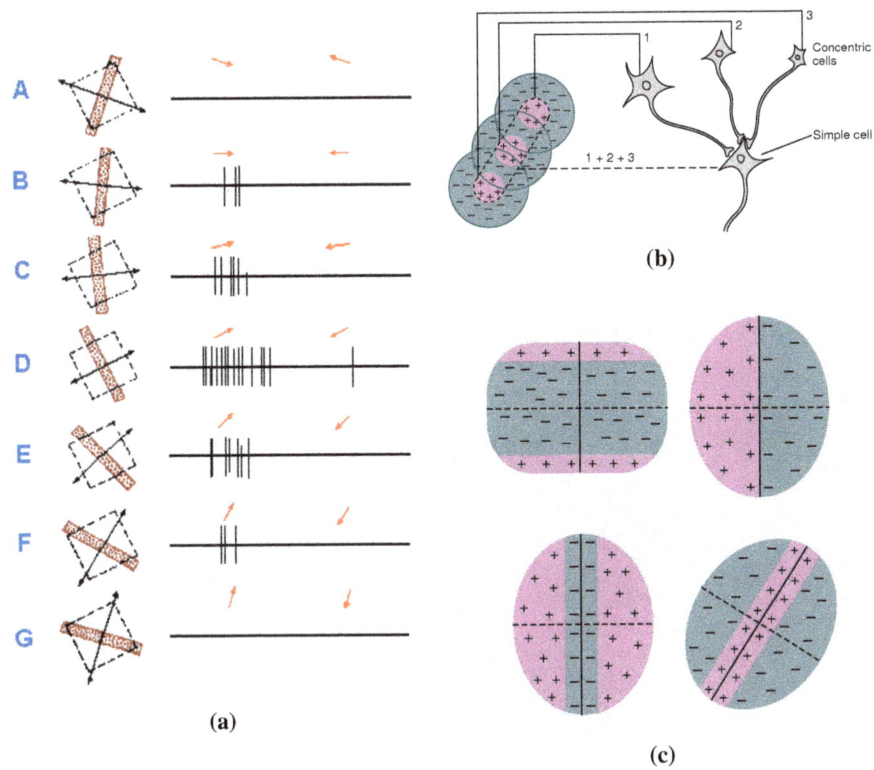

Fig. 2.4 V1 simple cell receptive fields. **(a)** Bars of different orientation (left) are shown to an anesthetized animal, and the firing patterns at a certain position in V1 are recorded by microelectrodes inserted into the cortex. The plots on the right show spikes elicited over time to the bars moving in the direction indicated by the little arrow. We can see that this group of V1 cells responds most strongly to a bar of orientation "D" moving to the right. (Reprinted from Hubel and Wiesel (1968) with permission of Wiley-Blackwell). This famous experiment by Hubel and Wiesel and subsequent ones have lead to today's textbook view as sketched in **(b)** and **(c)**: Several on-center LGN cells (called "concentric cells" here) project to a simple cell, resulting in an elongated RF similar to the bottom right one in **(c)**. Reprinted from (Kandel et al., 2000) with permission of McGraw-Hill.

and with a certain direction of movement (see Figure 2.4). Several simple cells again project to one complex cell, making it invariant of the exact position (i.e. phase) and movement direction of the edge. These properties of V1 neurons were first discovered in the groundbreaking experiments of Hubel and Wiesel (1968).

From V1, signals continue to V2 and then to V4. An important feature of V2 is the representation of *illusory contours* (first observed by von der Heydt et al., 1984). Illusory contours are contours perceived subjectively in a visual stimulus at positions where there is no such discontinuity or edge in the actual input (e.g. the Kanizsa triangle in Figure 2.5). Thus, the coding in V2 is an early step in a whole series that leads from veridical representation of the physical stimulus to a semantic representation of its perceived meaning. Note that the perception of illusory contours cannot be explained any more by pure feedforward processing. Instead, it requires intricate lateral interconnections within V2 and maybe feedback from higher areas.

Processing continues in V4, where larger, also interrupted, edges and corners made up of two edges are represented (Pasupathy and Connor, 1999). The ventral stream then reaches inferotemporal cortex (IT). While in the posterior IT, which is innervated by V4, most cells can be activated strongly by bars or disks just by adjusting the size, orientation, or color of the stimulus, cells in the subsequent central and anterior parts require more complex features for maximal activation like parts of whole objects. Finally, visual signals are combined with other signal streams in the superior temporal polysensory area (STP) to allow full recognition and

Fig. 2.5 Some V2 cells respond to illusory contours like the edges of the illusory white triangle (Kanizsa, 1955). While a V1 cell with preferred position somewhere along the contour of this triangle will not fire, a V2 cell with adequate preferred orientation will become active.

understanding of objects. There are additional dedicated brain areas for recognition and representation of important classes of objects (cf., e.g., Section 3.2).

2.1.1 Topographic Organization of the Visual Areas

Investigations of the properties of neighboring cortical neurons have revealed the *columnar* organization of cortex: In primary visual cortex, for example, all neurons within a column of roughly 1mm diameter receive input from one point in visual space. Their responses differ, however, in orientation of spatial frequency tuning. We will discuss the columnar organization of cortex in detail in Sect. 3.3.

On a slightly larger scale, however, visual information in V1 is ordered retinotopically. That is, neighboring points in the visual field are represented by neighboring (on the above-column scale) neurons in V1. Visual patterns on the retina are mapped to conformal, topologically consistent patterns of activity in cortex (see Figure 2.6). Note that while neighborhood relations are preserved by the retino-cortical mapping, the geometric shape of the visual representation can change.

This retinotopic organization of visual information is repeated at higher stages of the ventral stream. We have seen in the previous paragraphs that in the succession of stages from retina through IT, the visual features represented by the respective area become larger and more relevant for understanding. They also become more context dependent and less directly stimulus driven, and exact spatial localization decreases. Therefore, topographic organization in higher areas like inferotemporal cortex becomes more difficult to assess, but even there it is still roughly retinotopic.

So in a nutshell, the primate ventral stream is a progression of retinotopic maps with neurons (or groups of neurons) representing more complex features and larger regions the higher we come. One might think that this hierarchy is driven by pure feedforward flow of information. The opposite is true, however. Many studies have shown that, depending on the cortical region, up to 95% of cortical connections are feedback, that is they transmit information from a "higher" cortical area to a "lower" one. While the exact role of these feedback connections is still discussed, there is no doubt that higher cortical areas do influence the activity at lower areas. Murray et al. (2006), e.g., showed a two dimensional image of a ball to human subjects, and gave them additional information about the distance of the ball, thus influencing the subject's perception of the size of the ball. A larger perceived size resulted directly in a larger area in V1 representing the ball and vice versa, although the relevant bottom-up input to V1 had not changed at all. Thus, decisions at higher cortical areas can strongly influence the activity patterns in primary visual cortex. Similarly, lateral connections *within* a cortical area also play a strong role in visual information processing, as we saw above in the case of illusory contours in V2.

What is unknown and disputed, however, is how exactly the feedback and lateral connections influence cortical information processing. Do they simply modulate the feedforward flow of information? Or do they qualitatively change the way the bottom-up information is interpreted? Do they maybe even have some more important role like aligning and finding matches between the information represented

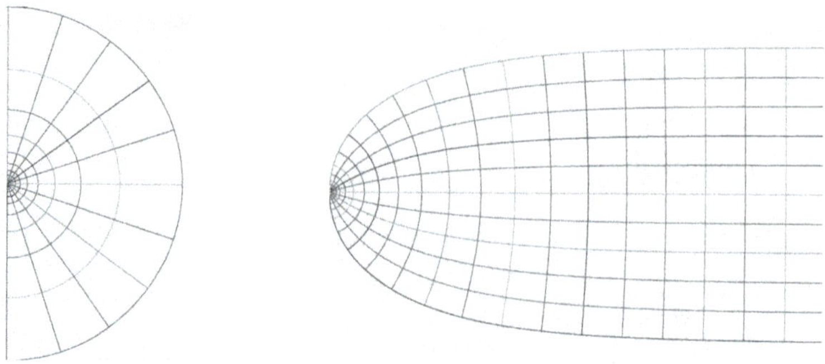

(a)

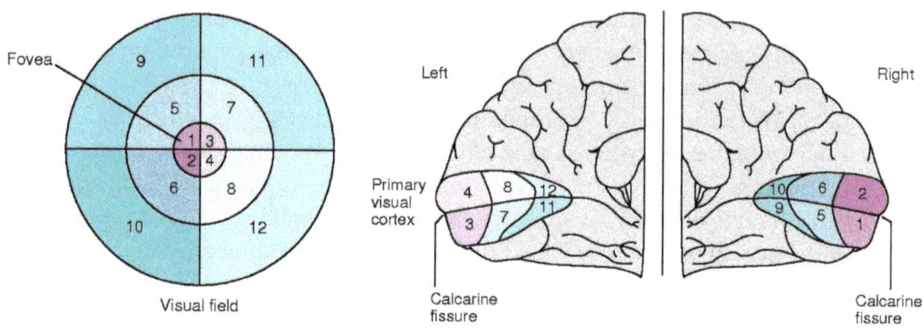

(b)

Fig. 2.6 Retinotopy. **(a)** Retinotopic representation in the Macaque V1. Left: Flickering stimulus shown to the animal. During the experiment, the contrast of the lines is reversed continuously to avoid adaptation of the animal. Right: Idealized sketch of the resulting pattern in V1 of the animal, as found by experiments like in Tootell et al. (1988). Dark lines—made visible by cytochrome oxidase staining—reveal the representation of the stimulus in V1. Note that while topological relations have been preserved, the geometric shape of the visual representation has changed: Concentric circles have changed to vertical lines, lines radiating from the center of the visual field have become horizontal lines. The center of the visual field is overrepresented compared to the periphery. This is the well known log-polar mapping from retina to cortex (see, e.g., Schwartz, 1977). **(b)** The corresponding text book interpretation as published in (Kandel et al., 2000), reprinted with permission of McGraw-Hill. The sub-parts of the visual field on the left are represented by the regions of V1 bearing the same numbers.

at the different levels? These questions have provided the motivation for most of the research presented in this book.

2.2 Approaches to Invariant Object Recognition

One issue that can help shed light onto this question is our ability for *invariant recognition*. If we see the same object under slightly changed circumstances, e.g. at a different position, our brain will recognize it as the same object as before without any problems, so smoothly in fact that in the early days of computer vision this was not even noticed as a serious problem. Yet the retinal pattern created by this new situation is entirely different. Its (mathematical) similarity with the original pattern may even be smaller than that of two patterns caused by different objects, but in the same position, as illustrated in Figure 2.7.

So how does our brain (and how can any computational system) solve this problem of recognizing the two images as being caused by the same object? The sheer amount of different situations in which we can recognize the same object makes it prohibitive to store all possible views in memory. If this is not possible, however, then our brain must have active mechanisms that recognize different patterns as coming from the same object. This is the problem of invariant recognition. By now,

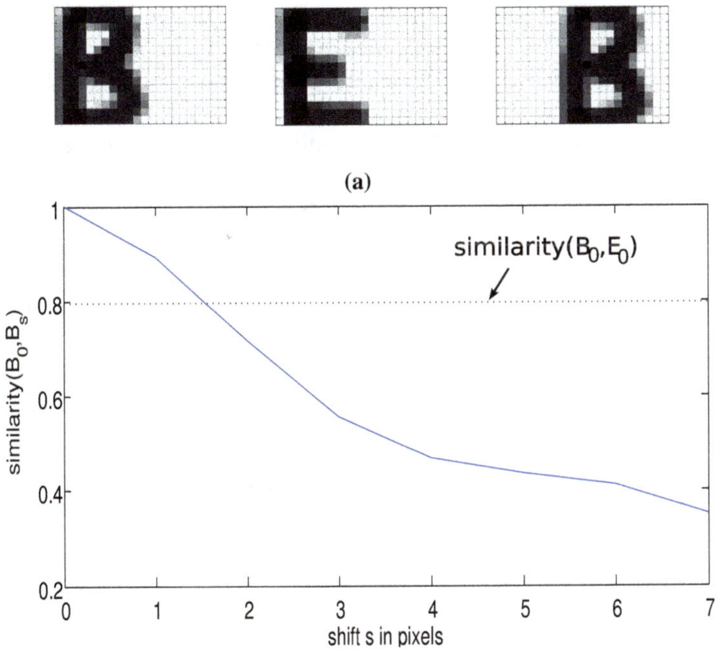

Fig. 2.7 Problem of invariance in machine vision. **(a)** shows grayscale pixel images of handwritten letters B and E, and on the right, the same letter B shifted by eight pixels. In **(b)**, the similarity—given by the dot product of the pixel vectors—of the letter E and the original letter B is indicated by a dashed line. The full line shows the similarity of B in its original position and a version of B shifted by s pixels. Already for a shift of 2 pixels, the similarity of the shifted versions of one and the same letter B is lower than the similarity of the different letters B and E. This demonstrates that straightforward similarity metrics like the dot product cannot enable position-invariant object recognition.

it has a history of more than 50 years of active research in such diverse disciplines as computer science and engineering, physics, neuroscience, and psychology, and it remains far from being solved. Over this period, a multitude of neural models has been proposed to explain invariant recognition. Although they all have their unique characteristics, they can roughly be cast into two different categories according to the underlying principles they follow.

2.2.1 Two Complimentary Concepts for Achieving Invariance

Pooling or feature-based approaches

A traditional approach in computational neuroscience, which we refer to here as the *feature-based* approach, aims to achieve invariance by converging the signals from feature detectors at different positions (and scale and orientation) in an earlier layer into a single "complex" unit at a higher layer. This idea was first proposed by Frank Rosenblatt in his four-layer perceptron (see Figure 2.8), and a multitude of similar models has followed since (Fukushima et al., 1983; LeCun et al., 1989; Mel, 1997; Riesenhuber and Poggio, 1999; Deco and Rolls, 2004). This convergence (also called *pooling*) of features at different positions, scales, etc., makes the response of the following complex unit invariant to those transformations. For example, a unit that pools over feature detectors at different positions will have a position invariant response. Feature hierarchies usually consist of several stages that on the one hand combine simple features into more and more complicated ones while at the same time pooling over increasingly large regions in transformational space to become more and more invariant. Both operations have to go hand in hand because in this approach there is an inherent trade-off between complexity of features and resolution at which they can be represented (cf. Serre et al., 2007).

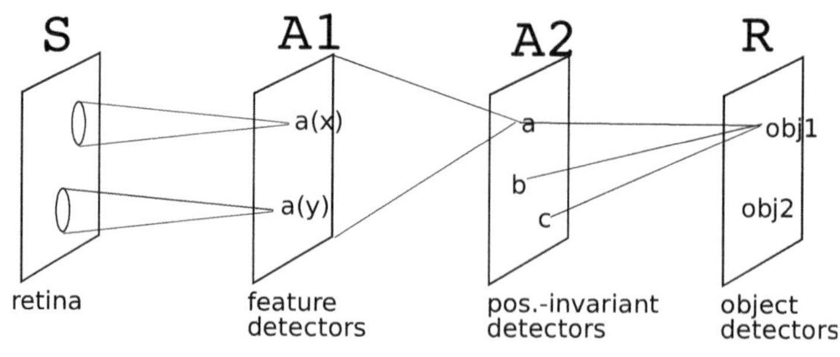

Fig. 2.8 The four-layer perceptron proposed by Rosenblatt (1961). It has a sensory surface S, two association layers $A1$ and $A2$, and a recognition layer R. $A1$ extracts local features from specific positions in S. $A2$ pools over all positions in $A1$ and thus reacts to features in the input invariantly of position. Units in R respond to certain combinations of $A2$ units that represent a specific object.

Correspondence-based approaches

The *correspondence-based* approach does not recognize objects by the activity of a single or a few cardinal cells, but through a pattern matching process that establishes explicit correspondences between points in the input image and parts of the object model stored in memory. The idea that this might happen via synchronization of patterns in the brain was proposed by von der Malsburg (1981) and developed successively into a neural model of correspondence finding (Wiskott and von der Malsburg, 1996). The idea of synchrony as a binding mechanism in the brain continues to be studied both experimentally (e.g. Gray and Singer, 1989; Singer, 2003) and theoretically (e.g. Wang, 2005). Olshausen et al. (1993) introduced the notion of dedicated *control units* to control the flow of information between patterns.

The principle of correspondence finding between patterns requires direct links or routing networks providing connections between the two patterns. Instead of pooling over (i.e. basically responding to any activity within) lower stages, the correspondence-based approach actively selects the information that is allowed to activate the next layer. This is done by dynamically switching connections between successive layers, thus routing input information to different parts of the next layer depending on the situation. Such a routing process can in principle compensate the effects of variances, rendering the information represented at the output stage invariant of the extrinsic properties of the input image like position, scale, pose, etc., without discarding information. Dynamic information routing was proposed as a principle of invariant perception already by Pitts and McCulloch (1947), and the term *dynamic link* was introduced by Feldman (1982). Several specific routing architectures have been proposed since (Anderson and Van Essen, 1987; Postma et al., 1997; Arathorn, 2002). Although the ideas of active information routing and correspondence finding have been studied mostly independently in the past, we will argue in this book that they mutually require each other in a full vision system.

2.2.2 *State of the Art in Object Recognition*

After this definition of the principal approaches in computational neuroscience, let us now review the performance of current systems for object recognition (both neural models and computer vision systems) in the light of these distinctions. Feature-based and correspondence-based models have been successful in different application domains, as discussed in (Biederman and Kalocsai, 1997). Feature-based approaches are very successful in classification tasks. The "standard model" from the Poggio lab (Riesenhuber and Poggio, 1999; Serre et al., 2007) is a refinement of the Neocognitron (Fukushima et al., 1983). It uses two successive stages of pooling and feature extraction and a support vector machine as a final classifier. Pooling operations (over different positions and scales) are hard-coded, while features are learned via weightsharing either simply as patches collected from training data or with a radial basis function approach. The model is competitive with state-of-the-art computer vision approaches (see below) in classifying objects like cars or airplanes, and shows some success in labeling parts of scenes as 'sky', 'building', etc. Other models in this direction include "LeNet" (LeCun et al., 1989, 2004) and the model of Deco and Rolls (2004). Wersing and Körner (2003) learn sparse features, and their highest nodes do not pool over the whole image, which reduces computational costs and allows distinguishing between more objects than the few object classes the above approaches can handle. Feature-based systems are suited for classification tasks because here their relative insensitivity to small metric variations of object parts is advantageous. Also, the static connectivity in feature-based systems allows them to be tuned to specific image databases. However, see (Pinto et al., 2008) for a discussion why such standard databases may not be a very good benchmark.

Correspondence-based systems, on the other hand, prevail in recognition tasks in which small differences in features and their arrangements are important. A typical example for this is face recognition. Since the very successful Elastic Graph Matching (EGM) model (Wiskott et al., 1997), the best performing systems for face recognition have mostly been correspondence-based (Phillips et al., 2003, see also www.frvt.org).

Let us now review how pure computer vision systems, whose only goal is to achieve the highest performance possible for certain tasks without following a specific "philosophy", can be classified according to the two above distinctions. The most obvious variance that a visual system has to deal with is translation of an object. While the most important biological mechanism for dealing with translations is eye movements (saccades), the origin of these saccades requires explanation in the first place. And even without eye movements we are capable—with some limitations under unnatural conditions (Cox et al., 2005)—to recognize images that have been shifted on the retina (Bar and Biederman, 1999; Fiser and Biederman, 2001).

When the (two-dimensional) Fourier transform is applied to an image, the resulting amplitude spectrum represents the global frequency content of the image and is therefore invariant to translations. This means that two images of the same object at different positions have the same Fourier spectrum. Pollen et al. (1971) suggested that this mechanism might be used in the visual system to achieve translation invariance, and it has been used to some extent in artificial vision systems. Unfortunately, the amplitude spectrum of the Fourier transform is not unique: since phase information is not retained, for any given image there are totally different, nonsensical images that have the same amplitude spectrum.

A generalization of the Fourier amplitude spectrum is the use of histograms that represent the number of certain features in an image without paying attention to spatial relations. Histogram approaches have a long history in computer vision (Schiele and Crowley, 2000; Swain and Ballard, 1991). The currently most popular approach in this direction are *bag-of-features* models, a name derived from similar *bag-of-words* approaches in document analysis (Joachims, 1998). Bag-of-features models (e.g., Leung and Malik, 2001; Lazebnik et al., 2003; Csurka et al., 2004) represent images as an unstructured set of image patches or other features. Since they do not model any spatial relations, only statistics of an image, they have been especially successful in scene classification (i.e. catching the gist of an image, like whether it shows an office environment, a street scene, or mountains).

Examples of work in this direction include (Oliva and Torralba, 2006; Torralba et al., 2003; Lazebnik et al., 2006). Pure bag-of-features models are related to the feature-based approach in its crudest form. Pyramid approaches with bags of features (Lazebnik et al., 2006) introduce a bit of spatial ordering to the features and correspond to multilayer feature hierarchies like (Fukushima et al., 1983) and subsequent models.

Although the simplicity of use and robustness to occlusions of bag-of-features models make them very popular in computer vision, tasks like object localization or accurate recognition usually require modeling of the geometric relations between object parts. One step in this direction is for example adding pairwise relations between neighboring features to the bag-of-features representation (Sivic et al., 2005). This approach is similar to the way neighborhood relations are encouraged in Elastic Graph Matching (Wiskott et al., 1997), although a full object model is still missing. Approaches sporting a full geometric object model include the generative models of Perona's group (Fei-Fei et al., 2003; Fergus et al., 2003; Song et al., 2003) or the geometric correspondence search of (Berg et al., 2005). In (Song et al., 2003), for example, human body shapes are represented by a mixture of decomposable triangulated graphs. Such a body model can be learned from unlabeled or labeled data and can then be used to detect moving humans in image sequences. Representing objects by flexible graph structures is exactly the approach taken by correspondence-based systems for face recognition as in (Wiskott et al., 1997).

Another question besides spatial representation of objects is how the choice of features helps achieve invariance. In this regard, feature-based and correspondence-based approaches in neural modeling differ. In the former approach, a feature hierarchy produces invariant features that can be used for classification. Since information about the original *variance* is discarded on the way, the recognition process cannot ensure any more that the features extracted from different parts of the image are actually consistent (see Section 2.3 for a further discussion). In the correspondence-based approach, it is the matching process that makes recognition invariant and simultaneously transforms non-invariant into normalized features. Since this matching process is global, it automatically ensures that invariances are globally consistent across the whole object. While it would go beyond the scope of this introduction to review the many kinds of feature types that are used in computer vision, let us focus on one specific type that is interesting with respect to the above distinction. The SIFT (*scale invariant feature transform*) extracts features that are scale and orientation invariant *without* discarding information about these variances. Keypoints are chosen by detecting extrema in scale-space (i.e. over position and different scales), and at those keypoints a local orientation is calculated on the basis of the local image gradient direction. Thus, every keypoint gets assigned a specific scale and orientation which is used subsequently to normalize local image information, yielding invariant local features. At the same time, the information about keypoint position, scale, and orientation can be used to ensure that the object recognition process uses only features which are mutually consistent in terms of their variances.

With rising computing power, probabilistic approaches to computer vision have received increasing attention in recent years. Factor graphs (Kschischang et al.,

2001) can be used to make very fast inferences about visual scenes, while generative models (e.g., Murray and Kreutz-Delgado, 2007) represent objects in explicit models including possible variances, enabling them to *generate* images of specific instances of an object. This is very similar to the way objects are represented by correspondence-based models, at least newer ones like the system developed in Chapter 3 of this book. Much effort in probabilistic modeling goes into how inference on them is carried out. Since many classic inference techniques are prohibitive owing to the sheer size of vision problems, correspondence-based neural models might actually provide inspiration here. For overviews of probabilistic approaches see (Yuille and Kersten, 2006; Chater et al., 2006).

2.3 Computational and Biological Plausibility of the Two Approaches

2.3.1 Computational Arguments

So what are the computational differences between feature-based and correspondence-based approaches to vision, and what consequences do they have in terms of performance of the resulting models? As we have seen above, a main difference between the concepts is whether they explicitly represent the spatial layout of objects. Feature-based approaches, which more or less neglect this information, consider an object as recognized when all its constituing features are present somewhere in the scene. This approach is perfectly fine for problems where the spatial arrangement of parts is more or less irrelevant, like catching the gist of simple scenes (those in Figure 2.9a are already too challenging, see below). In these cases, feature-based approaches are actually superior since their simplicity and their static connectivity make them very easy to train and optimize, while finding useful geometric scene models as correspondence-based systems would use them might turn out difficult. Nevertheless, even for recognizing scenes, some spatial information may be helpful. Bosch et al. (2008) show that scenes which are misclassified by pure feature-based approaches (as in Figure 2.9a) can be classified correctly with a system that combines discriminative (i.e. feature-based) approaches with explicit spatial models.

For real object recognition, however, this lack of spatial ordering is a serious disadvantage, since it makes a system susceptible to falsely recognizing as an object an image that contains the parts of this object, but in a completely scrambled setup (e.g. the scrambled face in Figure 2.9b). This problem is especially likely to occur in scenes with complex background, where the system might pick features present in the background to "hallucinate" an object. Newer models have solved this problem partially in two ways. Interleaving many of the feature extraction and pooling stages and limiting the range of pooling at any single stage can reduce the insensitivity to spatial arrangement of features to some extent (Serre et al., 2007). The other approach to alleviating the problem is using overcomplete dictionaries of features that are dedicated to specific object classes (Mel and Fiser, 2000). The hope is that this

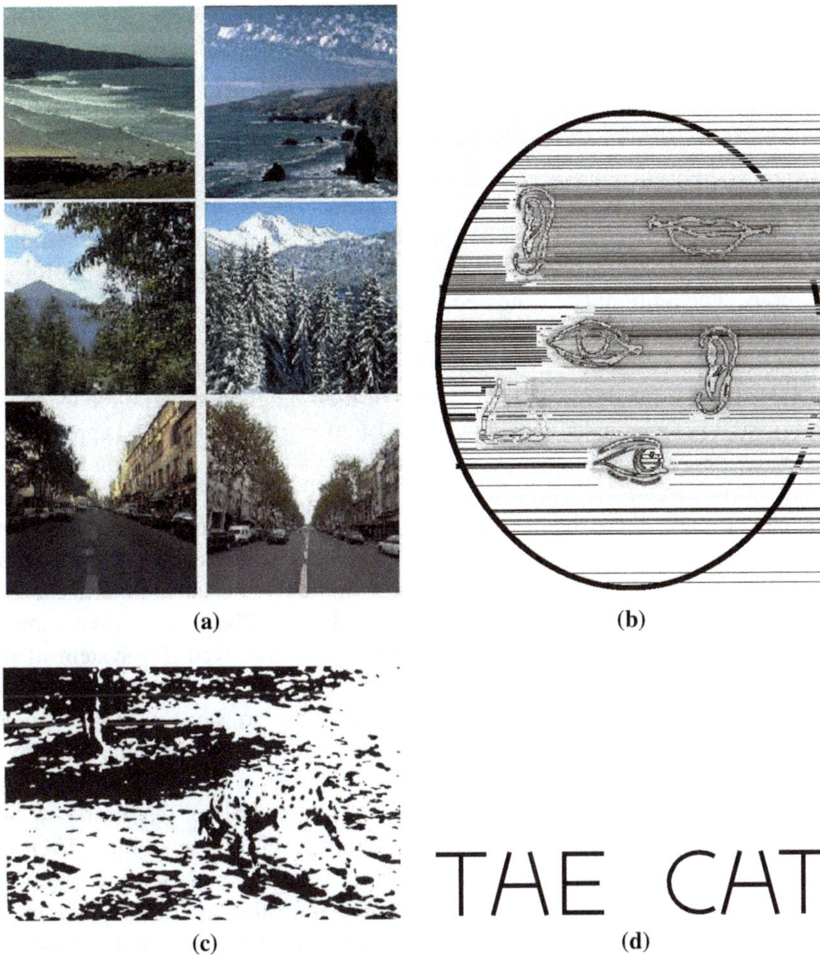

Fig. 2.9 These images illustrate situations that a vision system may encounter. **(a)** Landscape scenes. Without using spatial information, state-of-the-art feature-based approaches classify these coast and forest images as "mountain", and the street images as "highway". Images reprinted from (Bosch et al., 2008) with the permission of IEEE. **(b)** An image consisting of the scrambled parts of a face. **(c)** A Dalmatian. If you have ever seen this image, you will recognize it immediately. If you have not seen it before: the dog is in the right half of the image, walking towards the left, its muzzle to the ground. **(d)** What is written here?

will provide additional features that are sensitive to the spatial constellation of parts and can thus deal with scrambled images or background effects.

Nevertheless, there are visual tasks that require *exact* spatial information. An impressive one for example is our ability to recognize three-dimensional shapes in random dot stereograms. This requires that exact geometric correspondences be found, this time not between an image and internal memory, but between different regions of the input image. The single local features, random dots, are of no great value in finding these correspondences, rather a matching process between large constellations of points is required. Consequently, models addressing this task (e.g. Marr and Poggio, 1976) are correspondence-based. Most probably, any kind of stereopsis skill will require correspondence finding mechanisms.

A related question is the role that local and global decisions play in recognition. When looking at the famous Dalmatian in Figure 2.9c, for example, local features are totally useless, and only a global, model-driven recognition process can make sense of the image. In Figure 2.9d, on the other hand, local features are useful, but ambiguous. The ambiguity of the central letters in the two words can only be resolved by contextual feedback from the global decision. The classical feature-based paradigm does not support the notion of local feature detectors incorporating cues from global decisions or from the decisions of their neighbors, while feedback and local interactions are fundamental principles of correspondence-based approaches.

Leaving the question of spatial representation and global interaction aside, one problem of feature hierarchies remains: by pooling over variances, they do not only become invariant to them, but they effectively discard information about these variances! In consequence, there is no way of ensuring that features in the image assumed to represent a certain object are actually mutually consistent in terms of their variances. Again, overcomplete coding (Mel and Fiser, 2000) may solve some of these problems by introducing overlapping features. But even if a system of this kind is able to detect and recognize objects, it has no access anymore to the variances discarded along the hierarchy. For example, it has no way of telling where the recognized object is, what size it has, whether a person in the image has a happy or a sad expression on her face, etc.

Moreover, because the pooling operation is not invertible, feature-based systems cannot *generate* specific instances from high-level representations. Their object models are mere *detectors* instead of explicit models representing objects in all their possible variances. It has been argued (cf. the "predictive coding" of Rao and Ballard, 1999) that such capability to regenerate the current percept and compare it with the actual stimulus may be advantageous for a vision system, because it increases the signal-noise-ratio and allows a global consistency check of the features (compare the above discussion). And as mentioned above, the recent success of Bayesian models lies in their having explicit models of object appearances (the *likelihood* in Bayes' rule). It is explicit, generative models that enable advanced visual functions like reasoning about a percept, mental filling in of occluded regions, or testing hypotheses about it, in short that give us the feeling of being in direct contact with our visual environment. Correspondence-based systems do not automatically have explicit object models, but we will argue in this book how they can be implemented.

2.3.2 Experimental Evidence

Let us now review physiological, anatomical, and psychophysical evidence that argues for and against the two approaches. For this, it is interesting to look at what is known about feature processing and receptive fields (RFs) in the visual system. It is often argued that the primate ventral stream constitutes a kind of hierarchy of more and more complex features (Tanaka, 1996; Oram and Perret, 1994), from Gabor-like RFs in V1 to neurons in inferotemporal cortex (IT) that react invariantly to large parts of objects. This is exactly what feature-based systems like (Riesenhuber and Poggio, 1999) try to model. Correspondence-based systems, on the other hand, have relied so far on representing objects by groups of rather basic features, which appears less realistic. Note, however, that while V1 cells respond similarly in awake and anesthetized animals, it is nearly impossible to drive IT cells under anesthesia (for recent results on the large differences between general neuron responses in awake and anesthetized animals, see e.g. Greenberg et al. 2008). So the very complex effective RFs of IT cells in awake animals cannot directly correspond to anatomical RFs, since then they should respond similarly also under anesthesia. Instead, they might arise from interaction of many cells with simpler RFs. This is the way complex percepts like whole faces are represented in the model of Chapter 3. Nevertheless, correspondence-based models should try to address feature extraction more explicitly than in the past, especially since there are no fundamental obstacles for doing so.

The notion of effective RFs leads to the general question of how static or flexible RFs are. There is abundant physiological evidence that they are not static at all. Shifting receptive fields have been found in lateral intraparietal cortex (Duhamel et al., 1992; Kusunoki and Goldberg, 2003), in MT (Womelsdorf et al., 2006), and even in V2 and V4 (Luck et al., 1997). Therefore it would be possible that effective receptive fields in the visual system change from one instance to the next to route and match the current stimulus of interest to representations in memory.

A main argument for feature-based feedforward recognition has been the processing speed of the human visual cortex. Thorpe and coworkers have shown (Thorpe, 1988; Thorpe et al., 1996) that humans can decide whether an image contains an animal or not in less than 150ms. In the area of face recognition, Debruille et al. (1998) found that event-related potentials (ERPs) in response to novel vs. known faces start to differ as early as 76 to 130ms. Since such times are not much longer than the time required for a first wave of spikes to travel through the ventral stream after presentation of an image, it has been argued that visual recognition must be feedforward. However, such an interpretation seems to capture only part of the story. For instance, population codes can increase the speed of information transmission. The average spike rate of large excitatorily coupled neuron populations can be read out on a timescale that is much faster than the average spike latency of their single constituing neurons (van Vreeswijk and Sompolinsky, 1998). Thus, networks that have such "high gain" connectivity can respond very sensitively to subtle and fast input changes (similar to the principle of criticality; Bak, 1996). Furthermore, independently of population coding, correspondence-based systems can react very fast if they are primed (see below), i.e. if their dynamic connections have already been prepared for a specific stimulus. This might be the case in simple classification tasks like in the experiments of Thorpe (1988). In such cases, even correspondence-based systems react in a feedforward way.

Another point that begs discussion is the important role of priming or congruency effects in general object perception (for a review, see Graf, 2006). For example, when we look at the moon rising over a distant skyline, it looks much larger than when we see it high in the sky. This is because the size of the buildings around the moon primes our visual system for a certain scaling which is then unconsciously applied to the whole scene. Psychophysical experiments have shown that priming certain scales (Bundesen and Larsen, 1975) or orientations (Jolicoeur, 1985; Lawson and Jolicoeur, 1999) changes our performance and reaction times in object recognition. From these findings we can conclude that it does take effort and time to align the external world with internal representations, suggesting active dynamic processes for correspondence finding rather than passive pooling operations. In Section 3.5.4 we show how priming effects can be incorporated into correspondence-based models of object recognition.

Combining the evidence for feedforward processes on the one hand and correspondence-based ones on the other, it appears likely that the brain employs both strategies. This could be in the form that first there is a fast and unconscious feedforward sweep that is followed by more in-depth recurrent processing, only the latter leading to conscious perception (Lamme, 2003). This is consistent with findings by Johnson and Olshausen (2003), who report two ERP signals related to object recognition, an early presentation-locked one, and a later signal that correlates in timing with the response times for recognition.

2.4 Proposal for Dynamic Routing as a Principle of Brain Function

From the preceding discussion of computational, biological, and behavioral aspects we can conclude that the human visual system and most probably any other powerful object recognition system has to combine both feature-based and correspondence-based mechanisms. Wherever possible, the brain will employ feedforward mechanisms, since they are fast and undemanding. This may suffice in highly trained situations where immediate and stereotypical responses are required. Correspondence-based mechanisms, on the other hand, become necessary in ambiguous or novel situations, as well as in visual tasks beyond pure classification, like filling in of missing information or reasoning about a percept. As argued at the beginning of this chapter, we are convinced that dynamic information routing and the finding of correspondences also play a crucial role in other areas of brain function like auditory perception, understanding and producing syntactically correct speech, action planning, and producing appropriate motor outputs.

The necessity of correspondence-based mechanisms has been neglected in large parts of the neural modeling literature. In computer vision, on the other hand, it is well accepted, sometimes only showing up as an underlying principle, sometimes explicitly like in generative models and Bayesian approaches. Here, in turn, there have been few efforts to explain how the correspondence finding process could be implemented in a dynamic distributed system like the brain. This neurally plausible realization of correspondence finding processes will be at the focus of this book.

Even beyond vision, the general task of mapping corresponding patterns to each other (and, on the other hand, recognizing when two patterns do not match) is central to our survival and our intelligence. Although we seem to solve it without difficulties, this capability poses certain requirements to the brain as a physical and computational system:

1. Signal pathways must exist in the brain that allow routing of information between different parts of the brain, such that the patterns residing there can be compared. These pathways have to be manifold enough to allow the many types of routing and transformations we hinted at above; on the other hand they should be as parsimonious as possible for evolutionary reasons.
2. The brain must have computational mechanisms that implement the actual correspondence finding using these pathways. Realistically, this process has to be structured into several stages, to achieve high flexibility without drowning in a combinatorial explosion. These stages have to cooperate since the task of one alone can usually not be carried out without solving the full problem.

As we discussed already in the introduction, these requirements give rise to a multitude of questions: How can correspondence finding be implemented in the brain? What kinds of information routing pathways are advantageous? How can such structures self-organize during ontogenesis? And how can all this be integrated into a functional object recognition system? This book will hopefully contribute to some of those open questions.

References

Anderson, C.H., Van Essen, D.C.: Shifter circuits: A computational strategy for dynamic aspects of visual processing. Proceedings of the National Academy of Sciences of the United States of America 84, 6297–6301 (1987)

Arathorn, D.W.: Map-Seeking Circuits in Visual Cognition—A Computational Mechanism for Biological and Machine Vision. Stanford University Press, Stanford (2002)

Bak, P.: How nature works: the science of self-organized criticality. Springer, Heidelberg (1996)

Bar, M., Biederman, I.: Localizing the cortical region mediating visual awareness of object identity. PNAS 96, 1790–179 (1999)

Berg, A.C., Berg, T.L., Malik, J.: Shape matching and object recognition using low distortion correspondence. In: Proc. CVPR, pp. 26–33 (2005)

Biederman, I., Kalocsai, P.: Neurocomputational bases of object and face recognition. Phil. Trans. Roy. Soc. B 352, 1203–1219 (1997)

Bosch, A., Zisserman, A., Muoz, X.: Scene classification using a hybrid generative/discriminative approach. IEEE Trans Pattern Anal Mach Intell (PAMI) 30, 712–727 (2008)

Bundesen, C., Larsen, A.: Visual transformation of size. J. Exp. Psychol. Hum. Percept Perform 1(3), 214–220 (1975)

Chater, N., Tenenbaum, J.B., Yuille, A.: Probabilistic models of cognition: Conceptual foundations. Trends in Cognitive Sciences 10(7), 287–291 (2006)

Cox, D., Meier, P., Oertelt, N., DiCarlo, J.J.: Breaking position-invariant object recognition. Nature Neuroscience 8(9), 1145–1147 (2005)

Csurka, G., Dance, C.R., Fan, L., Willamowski, J., Bray, C.: Visual categorization with bags of keypoints. In: Proc. ECCV workshop on Statistical Learning in Computer Vision, pp. 59–74 (2004)

Debruille, J.B., Guillem, F., Renault, B.: Erps and chronometry of face recognition: following-up seeck et al. and george et al. Neuroreport 9(15), 3349–3353 (1998)

Deco, G., Rolls, E.T.: A neurodynamical cortical model of visual attention and invariant object recognition. Vision Research 44(6), 621–642 (2004)

Duhamel, J.R., Colby, C.L., Goldberg, M.E.: The updating of the representation of visual space in parietal cortex by intended eye movements. Science 255(5040), 90–92 (1992)

Fei-Fei, L., Fergus, R., Perona, P.: A bayesian approach to unsupervised one-shot learning of object categories. In: Proc. of the Ninth IEEE Intern. Conf. Computer Vision, pp. 1134–1141 (2003)

Feldman, J.A.: Dynamic connections in neural networks. Biol. Cybern. 46(1), 27–39 (1982)

Fergus, R., Perona, P., Zisserman, A.: Object class recognition by unsupervised scale-invariant learning. In: Proc. IEEE Conf. Computer Vision and Pattern Recognition (2003)

Fiser, J., Biederman, I.: Invariance of long-term visual priming to scale, reflection, translation, and hemisphere. Vision Research 41, 221–234 (2001)

Fukushima, K., Miyake, S., Ito, T.: Neocognitron: A neural network model for a mechanism of visual pattern recognition. IEEE Transactions on Systems, Man and Cybernetics 13(5), 826–834 (1983)

Graf, M.: Coordinate transformations in object recognition. Psychol Bull 132, 920–945 (2006)

Gray, C.M., Singer, W.: Stimulus-specific neuronal oscillations in orientation columns of cat visual cortex. Proceedings of the National Academy of Sciences of the USA 86(5), 1698–1702 (1989)

Greenberg, D.S., Houweling, A.R., Kerr, J.N.D.: Population imaging of ongoing neuronal activity in the visual cortex of awake rats. Nature Neuroscience (2008), `http://dx.doi.org/10.1038/nn.2140`

Hubel, D.H., Wiesel, T.N.: Receptive fields and functional architecture of monkey striate cortex. J. Physiol (Lond.) 195, 215–243 (1968)

Joachims, T.: Text categorization with support vector machines: Learning with many relevant features. In: Nédellec, C., Rouveirol, C. (eds.) ECML 1998. LNCS, vol. 1398, pp. 137–142. Springer, Heidelberg (1998)

Johnson, J.S., Olshausen, B.A.: Timecourse of neural signatures of object recognition. J. Vis. 3(7), 499–512 (2003), `http://dx.doi.org/10:1167/3.7.4`

Jolicoeur, P.: The time to name disoriented natural objects. Mem. Cognit. 13(4), 289–303 (1985)

Kandel, E.R., Jessell, T.M., Schwartz, J.: Principles of Neural Science, 4th edn. McGraw Hill, New York (2000)

Kanizsa, G.: Margini quasi-percettivi in campi con stimolazione omogenea. Rivista di Psicologia 49, 7–30 (1955)

Konen, C.S., Kastner, S.: Two hierarchically organized neural systems for object information in human visual cortex. Nature Neuroscience 11(2), 224–231 (2008), `http://dx.doi.org/10.1038/nn2036`

Kschischang, F.R., Frey, B.J., Loelinger, H.-A.: Factor graphs and the sum-product algorithm. IEEE Transactions on Information Theory 47, 498–519 (2001)

Kusunoki, M., Goldberg, M.E.: The time course of perisaccadic receptive field shifts in the lateral intraparietal area of the monkey. J. Neurophysiol. 89(3), 1519–1527 (2003)

Lamme, V.: Why visual attention and awareness are different. Trends in Cognitive Sciences 7(1), 12–18 (2003)

Lawson, R., Jolicoeur, P.: The effect of prior experience on recognition thresholds for plane-disoriented pictures of familiar objects. Mem. Cognit. 27(4), 751–758 (1999)

Lazebnik, S., Schmid, C., Ponce, J.: Affine-invariant local descriptors and neighborhood statistics for texture recognition. In: Proc. ICCV, pp. 649–655 (2003)

Lazebnik, S., Schmid, C., Ponce, J.: Beyond bags of features: Spatial pyramid matching for recognizing natural scene categories. In: Proc. CVPR 2006, vol. 2, pp. 2169–2178 (2006)

LeCun, Y., Boser, B., Denker, J.S., Henderson, D., Howard, R.E., Hubbard, W., Jackel, L.D.: Backpropagation applied to handwritten zip code recognition. Neural Computation 1(4), 541–551 (1989)

LeCun, Y., Huang, F.J., Bottou, L.: Learning methods for generic object recognition with invariance to pose and lighting. In: CVPR, pp. 97–104. IEEE Computer Society, Los Alamitos (2004)

Leung, T., Malik, J.: Representing and recognizing the visual appearance of materials using three-dimensional textons. International Journal of Computer Vision 43, 29–44 (2001)

Luck, S.J., Chelazzi, L., Hillyard, S.A., Desimone, R.: Neural mechanisms of spatial selective attention in areas V1, V2, and V4 of macaque visual cortex. J Neurophysiol. 77(1), 24–42 (1997)

Marr, D., Poggio, T.: Cooperative computation of stereo disparity. Science 194(4262), 283–287 (1976)

Mel, B.W.: Seemore: Combining color, shape, and texture histogramming in a neurally inspired approach to visual object recognition. Neural Computation 9, 777–804 (1997)

Mel, B.W., Fiser, J.: Minimizing binding errors using learned conjunctive features. Neural Computation 12(4), 731–762 (2000)

Murray, J.F., Kreutz-Delgado, K.: Visual recognition and inference using dynamic overcomplete sparse learning. Neural Computation 19(9), 2301–2352 (2007), http://dx.doi.org/10.1162/neco.2007.19.9.2301

Murray, S.O., Boyaci, H., Kersten, D.: The representation of perceived angular size in human primary visual cortex. Nature Neuroscience 9, 429–434 (2006)

Oliva, A., Torralba, A.: Building the gist of a scene: the role of global image features in recognition. Prog Brain Res 155, 23–36 (2006) http://dx.doi.org/10.1016/S0079-6123(06)55002-2

Olshausen, B.A., Anderson, C.H., van Essen, D.C.: A neurobiological model of visual attention and invariant pattern recognition based on dynamic routing of information. Journal of Neuroscience 13(11), 4700–4719 (1993)

Oram, M.W., Perret, D.I.: Modeling visual recognition from neurobiological constraints. Neural Networks 7, 945–972 (1994)

Pasupathy, A., Connor, C.E.: Responses to contour features in macaque area V4. J Neurophysiol. 82(5), 2490–2502 (1999)

Phillips, P., Grother, P., Micheals, R., Blackburn, D., Tabassi, E., Bone, J.: Frvt, evaluation report, Technical Report 6965, NISTIR. (2003), http://www.frvt.org/

Pinto, N., Cox, D.D., Dicarlo, J.J.: Why is real-world visual object recognition hard? PLoS Computational Biology 4(1), e27+ (2008), http://dx.doi.org/10.1371/journal.pcbi.0040027

Pitts, W., McCulloch, W.S.: How we know universals: the perception of auditory and visual forms. Bulletin of Mathematical Biophysics 9, 127–147 (1947)

Pollen, D., Lee, J., Taylor, J.: How does the striate cortex begin the reconstruction of the visual world? Science 173, 74–77 (1971)

Postma, E., van den Herik, H., Hudson, P.: SCAN: A Scalable Model of Attentional Selection. Neural Netw. 10(6), 993–1015 (1997)

Rao, R.P., Ballard, D.H.: Predictive coding in the visual cortex: a functional interpretation of some extra-classical receptive-field effects. Nature Neuroscience 2(1), 79–87 (1999)

Riesenhuber, M., Poggio, T.: Hierarchical models of object recognition in cortex. Nature Neuroscience 2(11), 1019–1025 (1999)

Rosenblatt, F.: Principles of Neurodynamics: Perceptrons and the Theory of Brain Mechanisms. Spartan Books, Washington (1961)

Schiele, B., Crowley, J.: Recognition without correspondence using multidimensional receptive field histograms. International Journal of Computer Vision 36(1), 31–50 (2000)

Schwartz, E.L.: Spatial mapping in primate sensory projection: analytic structure and relevance to perception. Biological Cybernetics 25, 181–194 (1977)

Serre, T., Wolf, L., Bileschi, S., Riesenhuber, M., Poggio, T.: Robust object recognition with cortex-like mechanisms. IEEE Trans. Pattern Anal. Mach Intell. 29(3), 411–426 (2007) (Evaluation Studies)

Singer, W.: Synchronization, binding and expectancy. In: Arbib, M. (ed.) The Handbook of Brain Theory and Neural Networks, pp. 1136–1143. MIT Press, Cambridge (2003)

Sivic, J., Russell, B., Efros, A., Zisserman, A., Freeman, W.: Discovering objects and their location in images. In: Proc. ICCV 2005, pp. 370–377 (2005)

Song, Y., Goncalves, L., Perona, P.: Unsupervised learning of human motion. PAMI 25(25), 1–14 (2003)

Swain, M., Ballard, D.: Color indexing. International Journal of Computer Vision 7, 11–32 (1991)

Tanaka, K.: Inferotemporal cortex and object vision. Annu. Rev. Neurosci. 19, 109–139 (1996)

Thorpe, S.: Identification of rapidly presented images by the human visual system. Perception 17, A77 (1988)

Thorpe, S., Fize, D., Marlot, C.: Speed of processing in the human visual system. Nature 381(6582), 520–522 (1996)

Tootell, R.B., Silverman, M.S., Hamilton, S.L., Switkes, E., De Valois, R.L.: Functional anatomy of macaque striate cortex. V. Spatial frequency. J. Neurosci. 8, 1610–1624 (1988)

Torralba, A., Murphy, K.P., Freeman, W.T., Rubin, M.A.: Context-based vision system for place and object recognition. In: Proc. ICCV 2003, pp. 273–280 (2003)

van Vreeswijk, C., Sompolinsky, H.: Chaotic balanced state in a model of cortical circuits. Neural Computation 10, 1321–1372 (1998)

von der Heydt, R., Peterhans, E., Baumgartner, G.: Illusory contours and cortical neuron responses. Science 224, 1260–1262 (1984)

von der Malsburg, C.: The correlation theory of brain function, Internal report, 81-2, Max-Planck-Institut für Biophysikalische Chemie, Postfach 2841, 3400 Göttingen, FRG (1981); Reprinted in Domany, E., van Hemmen, J.L., Schulten, K. (eds.): Models of Neural Networks II, ch. 2, pp. 95–119. Springer, Berlin (1994)

Wang, D.: The time dimension for scene analysis. IEEE Transactions on Neural Networks 16(6), 1401–1426 (2005)

Wersing, H., Körner, E.: Learning optimized features for hierarchical models of invariant object recognition. Neural Computation 15(7), 1559–1588 (2003), http://dx.doi.org/10.1162/089976603321891800

Wiskott, L., Fellous, J.-M., Krüger, N., von der Malsburg, C.: Face recognition by elastic bunch graph matching. IEEE Trans. on Pattern Analysis and Machine Intelligence 19(7), 775–779 (1997), http://www.cnl.salk.edu/~wiskott/Abstracts/WisFelKrue97a.html http://www.cnl.salk.edu/ wiskott/Abstracts/WisFelKrue97a.html

Wiskott, L., von der Malsburg, C.: Face recognition by dynamic link matching. In: Sirosh, J., Miikkulainen, R., Choe, Y. (eds.) Lateral Interactions in the Cortex: Structure and Function, Austin, TX. Electronic book, vol. 11, The UTCS Neural Networks Research Group (1996), http://www.cs.utexas.edu/users/nn/web-pubs/htmlbook96/, http://www.cnl.salk.edu/~wiskott/Abstracts/WisMal96c.html

Womelsdorf, T., Anton-Erxleben, K., Pieper, F., Treue, S.: Dynamic shifts of visual receptive fields in cortical area MT by spatial attention. Nature Neuroscience 9(9), 1156–1160 (2006)

Yuille, A., Kersten, D.: Vision as Bayesian inference: analysis by synthesis? Trends in Cognitive Sciences 10(7), 301–308 (2006)

Chapter 3
A Correspondence-Based Neural Model for Face Recognition

In this chapter we develop a correspondence-based model for object recognition. We will focus here on the question how correspondence finding can be realized neurally, using very simple assumptions for the underlying routing structures (a more realistic treatment of these will be given in Chapter 4).

The general underlying principle of correspondence finding is similar to that developed in (Bienenstock and von der Malsburg, 1987). We introduce this principle in Section 3.1. The function of the system as a whole is related to elastic graph matching (EGM) systems (Lades et al., 1993; Wiskott et al., 1997). However, while EGM is an algorithmic system that explicitly minimizes energy functions to arrive at the final solution, the model proposed here is a biologically plausible network of cortical columns. And while EGM requires two subsequent steps of first object localization and then identification which do not interact, the present system integrates these steps into one coherent dynamic network, so that the outcome of both the localization and the recognition process is the final state of one large dynamic system.

The system was developed mostly with the application to face recognition in mind, a topic that we introduce and motivate in Section 3.2. The basic computational unit of the network is a model of the cortical column, which we introduce in Section 3.3. On this basis, we proceed to outline the full network in Section 3.4. We report the behavior of the network and test it for different tasks in Section 3.5, before concluding the chapter in Section 3.6. The results discussed in this chapter were partially published in (Wolfrum, Lücke and von der Malsburg, 2008) and (Wolfrum, Wolff, Lücke and von der Malsburg, 2008), the material presented in Section 3.5.4 in (Wolfrum and von der Malsburg, 2008). Figures 3.4a, 3.5a, 3.7, 3.9, 3.11, 3.12, and 3.13 were reprinted from (Wolfrum, Lücke and von der Malsburg, 2008) with the permission of ARVO.

3.1 Correspondence Finding

How do correspondence-based systems find correspondences between images in a distributed, not centrally organized way? The basic problem is illustrated in

P. Wolfrum: Info. Routing, Corres. Finding, and Object Recognition in the Brain, SCI 316, pp. 29–67.
springerlink.com © Springer-Verlag Berlin Heidelberg 2010

Figure 3.1a, which shows two stick-figures as input and model. Both of these objects are represented by a layer of *feature units* (black circles). The general correspondence problem is to relate points in the input image to their corresponding counterparts in the model image, making it possible to map one image onto the other. When those images are represented neurally, it means that corresponding units have to be identified and their relationship has to be marked somehow. We do this by activating *links* between the units. In Figure 3.1a, black lines represent active links (i.e. correct correspondences) as a subset of all potential connections (gray lines).

As a prerequisite for correspondence finding, feature similarities must be computed. In the case of Figure 3.1a, simply activating links between those units with the highest similarity solves the correspondence problem. Unfortunately, in realistic scenarios high feature similarity is not sufficient to find correct correspondences. Different images of the same object may vary greatly, leading to high similarity between non-corresponding points (see, e.g., Wiskott, 1999). Unrelated features in the background of an image may add to this confusion. Figure 3.1b shows this in cartoon form, heavy lines connecting the features with highest similarity. The interference of the background structure above the head of the stick figure and the changed appearance of neck and legs results in wrong or ambiguous correspondences in this case. For realistic inputs such situations are very frequent and the ambiguities increase the more kinds of feature detectors are used. For a human observer, in distinction, it is easy to find correct correspondences, also in Figure 3.1b. The reason for this is that an object is defined by its features *and* their spatial arrangement. Correspondence-based systems therefore also have to take both of these cues into account. We do this here by allowing *topologic interaction* between links (see Figure 3.1c). Links representing mutually consistent correspondences (parallel links in this simple case) strengthen each other, while mutually exclusive links (links emanating from the same node) inhibit each other. With the right balance between topologic interaction and feature similarity influence, this method will find the right global correspondences in spite of local feature discrepancies. This approach was first taken in dynamic link matching systems (Lades et al., 1993; Wiskott and von der Malsburg, 1996; Würtz, 1995). Here, we take the same principal approach, but use explicit units that control the connectivity between layers similarly to control units in shifter circuits (Olshausen et al., 1993). In (Lücke et al., 2008) a system is described that finds correspondences between two patterns using this approach.

3.2 Face Recognition

The object recognition system presented in this chapter was developed with a focus on and mainly applied to face recognition—textbf. Faces have a well-defined spatial layout, which allows them to be treated holistically (Biederman and Kalocsai, 1997), obviating the need to address hierarchical composition out of sub-objects (Biederman, 1987). On the other hand, the similarity of different faces in general appearance requires very fine discrimination concerning both the exact location of landmark points and textural differences. These two constraints (compact shape and

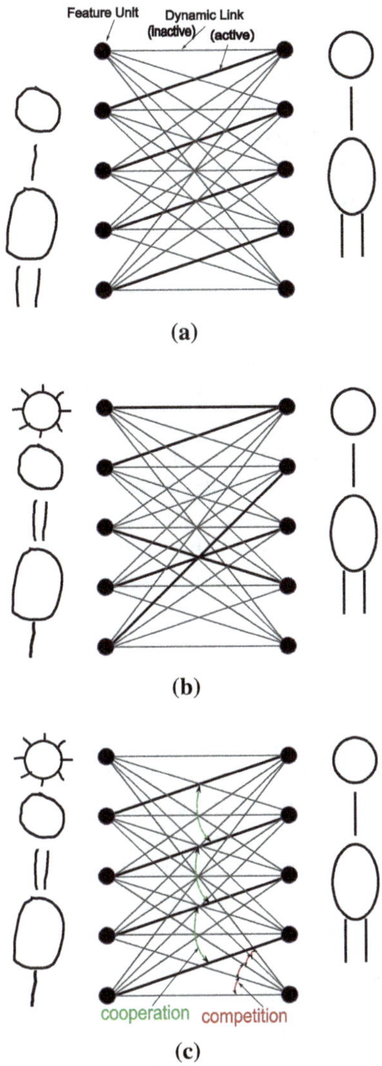

Fig. 3.1 The visual correspondence problem is the task of linking corresponding points between two images. **(a)** Input and model images are represented by arrays of feature nodes (black circles). All potential correspondences are symbolized by lines between the feature nodes. High feature similarities are indicated as heavy lines. In this case they represent the correct correspondence. In **(b)**, evaluation of feature similarity alone leads to wrong correspondences. **(c)** This problem is solved by additional interaction between dynamic links, which helps finding the correct global correspondence. Competition suppresses multiple matches to a single node, while cooperation encourages globally consistent mappings.

sensitivity to details) make face recognition specifically suited for correspondence-based approaches.

Face recognition is interesting because it is an important capability of the human mind, the ability to perceive and interpret faces being central to human social interactions. There is abundant psychophysical and pathological evidence that face recognition is a special case of human object recognition. For example, *prosopagnosia* patients are impaired at recognizing faces, while their other object recognition abilities are relatively normal. These patients can recognize a person by their voice or dress, but not from their facial image, indicating that there are dedicated neural processes responsible for identification of facial images. As another example, human infants are born with dedicated processing abilities for face-like visual patterns. They are more attracted by moving facial images than by images of other objects or jumbled facial features. Consequently, there exist dedicated neural resources for face recognition in the brain. While the fusiform face area (FFA) is specialized in face *recognition* (Kanwisher and Yovel, 2006; Tsao et al., 2006), neurons in medial frontal cortex seem to be involved in face *detection* (Summerfield et al., 2006), and the is responsible for evaluating *emotional* cues, mainly fear.

Face recognition has a well-established history in computer vision (see, e.g., Goldstein et al., 1971, who use a hand-designed vector of 21 features for face recognition). Differently to many other object recognition areas, stiff competitive tests are carried out on widely available image galleries (e.g., Phillips et al., 2000; Messer et al., 2004; Phillips et al., 2005). The existence of such tests and databases allows objective judgment of the performance of single systems. When photos are taken under controlled conditions, the performance of technical systems can be as good as that of humans or it even exceed it (Adler and Schuckers, 2007). In difficult situations, however, humans still outperform machine vision systems, encouraging us to investigate deeper the principles that lie behind human face perception. Thus, face recognition is a very relevant and mature field with much experimental background available. This nourishes the hope that developing a model of face recognition that on the one hand is guided by many of the relevant neurobiological facts and on the other hand comes close to the functional performance of state-of-the-art technical systems can help gaining genuine insight into the operating principles of the brain. This is what we set out to do in the following sections.

Apart from faces, there is evidence suggesting that FFA can also serve as an area of expertise (Tarr and Gauthier, 2000; Gauthier et al., 2000) for other object classes. In the same sense, our model is not confined to face recognition, but could be used for recognition of any kind of object type that has a prototypical shape and requires high sensitivity to small differences among objects.

3.3 The Basic Computational Units: Cortical Columns

3.3.1 Neurobiological Background

Our system for face recognition is implemented as a large network of cortical columns. The model we use to simulate the dynamics of a column is motivated

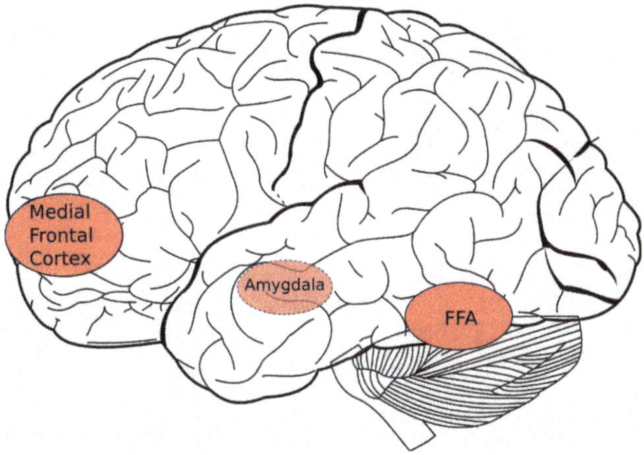

Fig. 3.2 Coarse sketch of the location of some cortical areas involved in face recognition. The medial frontal cortex is located around the superior frontal gyrus and the cingulate gyrus in the anterior part of the brain. Generally speaking, the frontal cortex is responsible for higher mental functions, action selection, and planning. The fusiform face area forms a part of the ventral stream and is located in the fusiform gyrus. It is defined functionally and can be localized in a subject by comparing fMRI (functional magnetic resonance imaging) or PET (positron emission tomography) responses to faces and other objects. The FFA is important for face recognition, but some studies (Tarr and Gauthier, 2000; Gauthier et al., 2000) suggest it can function as an area of expertise for any category of objects on which a person is highly trained. The amygdala, on the other hand, is clearly defined anatomically. It is a subcortical group of nuclei located close to the basal ganglia. The amygdala plays a prominent role in the processing of emotions. This is important for learning to recognize faces, since it is necessary to understand the social relevance of faces in order to pay enough attention to the minute details that distinguish different faces. Autistic children, who usually perform worse on tests of face and emotion recognition, also tend to avoid eye contact, and the amygdala of autistic subjects is often underdeveloped (Amaral et al., 2008).

by anatomical and physiological properties of the cortex on the scale of a few hundred microns. In particular, it reflects the columnar organization of the cortex (see, e.g., Mountcastle, 1997) and the concept of canonical cortical microcircuits as suggested, e.g., by Douglas et al. (1989). Columns are physiologically defined groups of neurons that extend through all cortical layers and have a diameter of roughly one mm. In some cases, they can be made visible through staining (Figure 3.3a). Depending on the perspective or the cortical area, a cortical column is commonly referred to as macrocolumn (Mountcastle, 1997), segregate (Favorov and Diamond, 1990), hypercolumnhypercolumn (Hubel and Wiesel, 1977) or simply column (e.g. Yoshimura et al., 2005). In primary visual cortex, a column comprises all neurons that receive input from one point in visual space.

The analysis of the fine-structure within a column suggests disjunct populations of excitatory neurons as functional elements. Anatomically, axons and dendrites of pyramidal cells have been found to bundle together and to extend orthogonally to

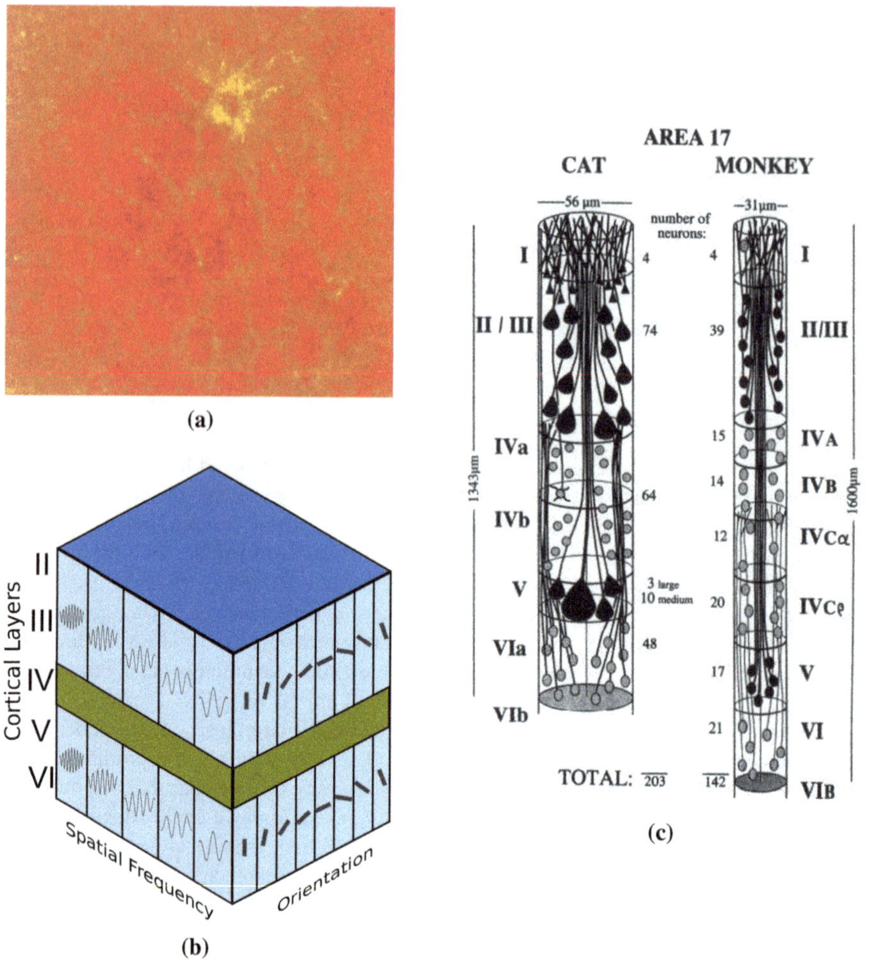

Fig. 3.3 Columnar organization of cortex. **(a)** Columns ("barrels") in rat barrel cortex, made visible through cytochrome oxidase staining. Reprinted from (Troncoso et al., 2004) with permission of Oxford University Press. **(b)** "Ice-cube"-like sketch of a hypercolumn in primary visual cortex. Each minicolumn represents a specific spatial frequency at a specific orientation. **(c)** Drawing of pyramidal cell modules in cat and monkey primary visual cortex. Reprinted from (Peters and Yilmaz, 1993) with permission of Oxford University Press.

the pial surface through the cortical layers. All neurons that directly contribute to one such bundle form a thin columnar module of just a few tens of microns in diameter (Peters and Yilmaz, 1993; Buxhoeveden and Casanova, 2002), as shown in Figure 3.3c. Together with associated inhibitory neurons (see, e.g., DeFelipe et al., 1989; Peters et al., 1997) such a module was termed *minicolumn* (Favorov and Kelly, 1994; Buxhoeveden and Casanova, 2002; Mountcastle, 2003) and was suggested as the basic computational unit of cortical processing (but see (Jones, 2000) or (Rockland and Ichinohe, 2004) for critical discussions). More recent evidence for disjunct functional units within a cortical column comes from experiments using focal uncaging of glutamate combined with intracellular recordings (Yoshimura et al., 2005). It was found that a column has a fine-structure of functionally relatively disjunct populations of layer 2/3 pyramidal cells. The relation of these populations to the cortical minicolumn has yet to be clarified, however. The main potential difference is that the concept of a minicolumn requires neurons in a population to be spatially adjacent whereas for neurons in the functional populations described in (Yoshimura et al., 2005) this is not necessarily the case.

Independent of the spatial arrangement of a column's functional sub-populations, there is little dispute about the existence of lateral coupling of such populations via a system of inhibitory neurons (Peters et al., 1997; Yoshimura et al., 2005). Yoshimura et al. (2005) for example have found the excitatory populations of layer 2/3 to receive common and population-unspecific input from inhibitory neurons of the same layer as well as from inhibitory neurons of layer 4 (see also Dantzker and Callaway, 2000).

3.3.2 A Model of the Cortical Column

We will define our dynamic model of a cortical column in accordance with these experimental findings. To be somewhat independent of different terminologies used in different communities, we will refer to the cortical column simply as *column* (instead of, e.g., macrocolumn or hypercolumn) and we will refer to its functional subpopulations as the column's *units*.

Generally speaking, a column represents all relevant features that are present at one location of either external, retinotopic space (cf. Figure 3.3b), or in some internal coordinate frame. Each unit of a column represents one such feature or quality. If necessary, competition among its units allows a column to represent only the strongest qualities at its location in a soft winner-take-all manner (see below). According with anatomical findings, each unit stands for an assembly of approximately 100 neurons. Since these neurons all represent the same feature, their mean firing rate (also called *population activity*) can be used to encode that feature. Contrary to using the average firing rate of a single neuron, however, this code is much faster and more reliable (for mean-field arguments see, e.g., (Wilson and Cowan, 1973; van Vreeswijk and Sompolinsky, 1998; Gerstner, 2000), and (Lücke and von der Malsburg, 2004) for a columnar model). This fast and robust information processing is a reason why our model can achieve recognition times comparable (in neural time scales) to human performance in spite of its inherently recurrent processing.

We describe the unit's neural activity by a differential equation called *modified evolution equation*. This equation represents our model of inhibition amongst the column's units and is a generalization of the well-known deterministic evolution equation (see, e.g., Eigen, 1971).

The activity x_i of the ith unit in a column of K units is given by

$$\tau \frac{d}{dt} x_i = x_i^{\nu} I_i - x_i \sum_{j=1}^{K} I_j x_j, \tag{3.1}$$

where τ is a time constant and I_i represents the input to unit x_i. The exponent ν parameterizes the competition strength among the units. This competition signal is global and changes during the recognition process (see below). However, it may be shifted in time for the different layers of the network, and it does not have an effect on all columns. In (Körner et al., 1999) the source of a fast and global modulatory signal to the cortex is discussed as the intralaminar nuclei of thalamus.

For $\nu = 0$, there is no competition, and (3.1) simplifies to

$$\tau \frac{d}{dt} x_i = I_i - x_i \sum_{j=1}^{K} I_j x_j. \tag{3.2}$$

In this case, all units represent their input proportionally, while the interaction term $\sum_j I_j x_j$ leads to activity normalization in the column (see Appendix A.1 for a proof). For $\nu = 1$, on the other hand, we get the dynamics

$$\tau \frac{d}{dt} x_i = x_i (I_i - \sum_{j=1}^{K} I_j x_j). \tag{3.3}$$

Now we have strong competition among the units, leading to winner-take-all (WTA) behavior (again, see Appendix A.1 for a proof and further analysis).

In our model of object recognition we assume that there are two types of columns with different functions. Dynamically, they only differ in the use of the competition parameter ν:

- *Feature columns* represent their input in a linear fashion (see Figure 3.4a). Consequently, the units in a feature column have no need to compete among each other, i.e. for them the parameter $\nu = 0$.
- *Decision columns* show a WTA behavior leading towards a state where only the unit getting the strongest input remains active. These units receive a ν-signal that linearly rises from 0 to 1[1]. So they start out with linear dynamics like feature

[1] In principle, the competition parameter ν could be set to a constant value of $\nu = 1$. However, slowly increasing competition within the columns of a network has in earlier systems proven to efficiently avoid local optima (Lücke et al., 2008). This is related to the slow change of the temperature parameter in simulated annealing like systems (Kirkpatrick et al., 1983), which serves the same purpose.

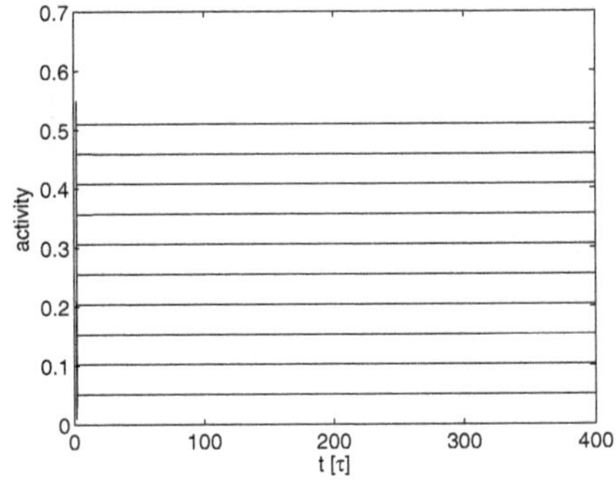

(a) Feature column ($\nu \equiv 0$).

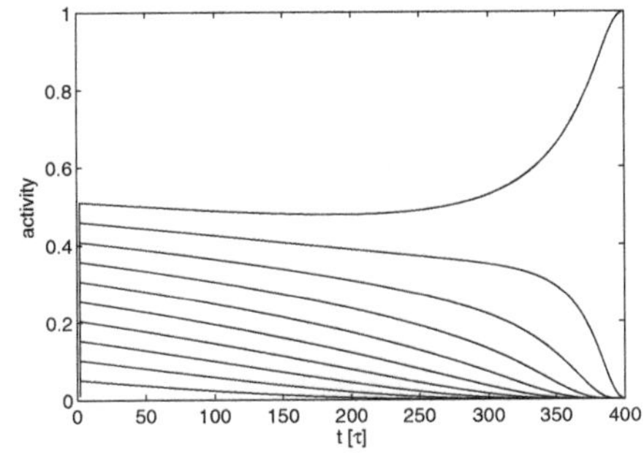

(b) Decision column (ν rises from 0 to 1 over the shown time period).

Fig. 3.4 Typical time course of the unit activities in an isolated feature column (a) and decision column (a). The inputs to the $K = 10$ units are spread equidistantly between 0 and 0.5. **(a)** After a sharp initial rise, a feature column represents its inputs in a linear fashion. **(b)** For a decision column, the competition parameter ν rises from 0 to 1 during the cycle time of $T = 400\tau$. The column starts out with activities proportional to the inputs like a feature column. Rising values if ν induce rising competition among the units, finally leaving only the unit with strongest input active. Note that the WTA behavior seen here results directly from the growth of the competition parameter ν. The internal dynamics of a column is much faster, so that with respect to the slow growth of ν, a column is always in quasi-steady-state. This can be seen also in the fast rise of the unit activities from very small initial values to the significantly higher steady states at the very start of the plot.

columns. With rising ν, competition increases, finally leading to a WTA behavior that leaves only the unit with the strongest input active. The typical dynamics of a decision unit is shown in Figure 3.4b.

The crucial computations in our system are performed by decision columns, whereas feature columns serve for information representation. Both kinds of columns may actually have the same neural substrate with the only difference that feature columns do not receive (or just do not respond to) the ν signal.

In the networks that we will introduce in the following section, units communicate with units of other columns. For this communication, a column scales the output activities of its K units such that its output energy (i.e. the Euclidean norm of the column activity vector) stays constant[2]:

$$\bar{x}_i := \frac{x_i}{\sqrt{\sum_{j=1}^{k} x_j^2}}. \tag{3.4}$$

[2] For brevity of notation, we will sometimes just use the name of a certain unit type (like $\mathscr{C}$ for control units) to denote the output of that unit. We will always point this out when we do so.

This kind of output normalization is advantageous for maintaining homeostasis in networks of columns and may be carried out by neurons in layer 5 of the cortex as suggested by Douglas and Martin (2004). Note that for feature columns this Euclidean normalization happens automatically in steady state (cf. Appendix A.1). For decision columns, explicit normalization is only necessary during the central phase of the cycle. At the beginning, it follows feature column dynamics anyway, while activity in the final state has both a 1-norm and a 2-norm of 1.

3.4 The Network

The principal architecture of the system is roughly visualized in Figure 3.5a. It consists of three main parts, an Input Layer for image representation, an Assembly Layer, and a Gallery Layer as memory. The Assembly Layer establishes correspondences between input and memory. It recurrently integrates information about feature similarity, feature arrangement, and face identity. Given an input, the integration of these information components results in the system to converge to a state that represents a percept. Figure 3.5a sketches the system after such a convergence when it has correctly established correspondences between a person's face stored in memory (i.e. in the Gallery Layer) and a given input image of this person. The principle of information integration from both the Input and the Gallery Layer in the Assembly Layer is sketched in Figure 3.5b. Note the inherent symmetry of bottom-up and top-down information flow (however, as we will see below, this information flow is realized in different ways). In the following, we will discuss the architecture of the system in detail.

As we saw before, the largest subunits of the network are *layers*. These loosely correspond to the different cortical areas that make up the visual system (we are not speaking here of the layers of anatomically different neurons that can be distinguished within one area of cortex). Layers are organized topologically, with a topology that may be stimulus space, like in V1 and somatosensory cortex, or a more abstract space. The layers of our network interact recurrently and activity collectively converges towards a final state that represents the "percept" of the network, in our case the possible recognition of a face.

Layers may contain both feature columns and decision columns. If we assume every feature column to represent all relevant features at one position of a retinal image, then layers of feature columns can represent whole images. The network introduced below uses layers of two different spatial arrangements:

- Rectangular grid: Straightforward representation suitable for any image. Every column represents one specific geometric location (see Figure 3.6a).
- Face graph structure: An arrangement specifically suited for faces, where each column represents an important landmark position on a face (Figure 3.6b). Note that in this case, a column does not necessarily represent a fixed spatial location in the image, but rather a fixed semantic location (nose, mouth, eye, chin, etc.). Spatial locations of landmarks can change according to the face they represent.

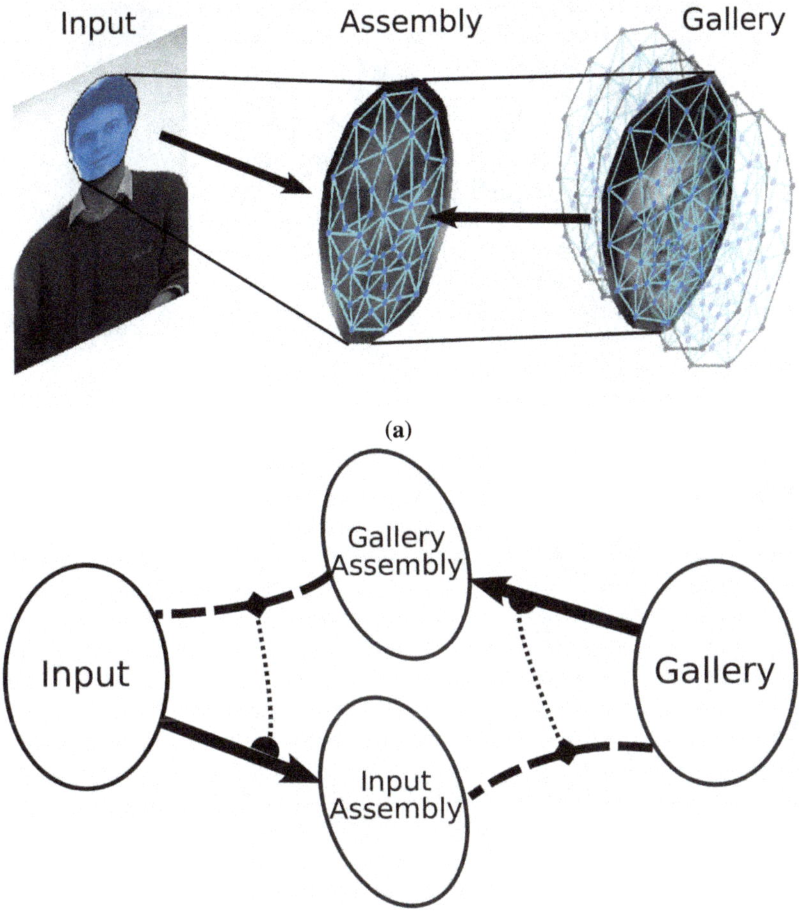

(b)

Fig. 3.5 Principal layout of the system. **(a)** The system has to simultaneously represent information about position *and* identity of the input face and its parts. Positional information is represented by dynamic links establishing correspondences between points in the input image and an in the internal reference frame ("Assembly Layer"). Identity information is represented by the activity of Gallery units, different graphs storing memories of different faces. **(b)** Both modalities contribute to the activity of the internal Assembly Layer, which represents visual information in its two sublayers Input Assembly and Gallery Assembly. Information flow to the Input Assembly is controlled by correspondences between Input and Gallery Assembly, while information flow from the Gallery to the Gallery Assembly depends on the similarity of the Input Assembly and models stored in the Gallery.

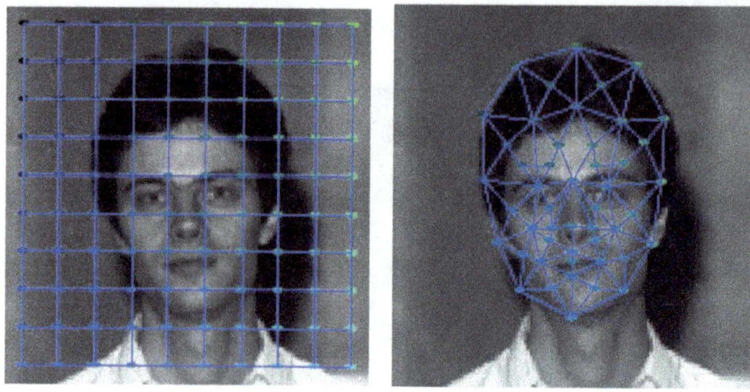

(a) Rectangular grid **(b)** Face graph

Fig. 3.6 Different ways of representing facial images. A rectangular grid graph (a) is used for general input image representation. Since in this case the positions of the relevant parts of the image are not known a priori, it is essential to represent the full image information by dense, homogeneous sampling. In cases where these positions are known, like for internal representation of faces, a face graph (b) consisting of characteristic points (landmarks) is a better data structure, since it is more parsimonious and allows pointwise comparison of images (see text).

The network consists of the following three layers (see Figure 3.7):

- Input Layer $\mathscr{I}$: Represents the input image in a rectangular grid.
- Assembly Layer: Integrates intermediate information from both the input image (represented in the *Input Assembly* units $\mathscr{IA}$, see Figure 3.9) and the gallery (represented by the *Gallery Assembly* units $\mathscr{GA}$).
- Gallery Layer $\mathscr{G}$: Represents all gallery faces in terms of the weights of its afferent and efferent connections to the Assembly Layer.

The following three subsections describe these layers in detail.

3.4.1 Input Layer

The Input Layer represents the input image using 400 feature columns arranged in a rectangular grid of $P = 20 \times 20$ points. Each feature column represents by its units' activities K features extracted from the image at that position.

If we neglect color and binocularity, the response properties of neurons in primary visual cortex are commonly described by the well-known Gabor wavelets (Ringach, 2002; Jones and Palmer, 1987; Daugman, 1980). In our model we use a predefined set of Gabor wavelets that appropriately sample orientation (over 8 orientations) and spatial frequency (over 5 scales) space, resulting in a number of $K = 40$ features at each point. That is, we use Gabor filter responses to model the RFs of the feature units in the Input Layer. For extracting the filter responses, we

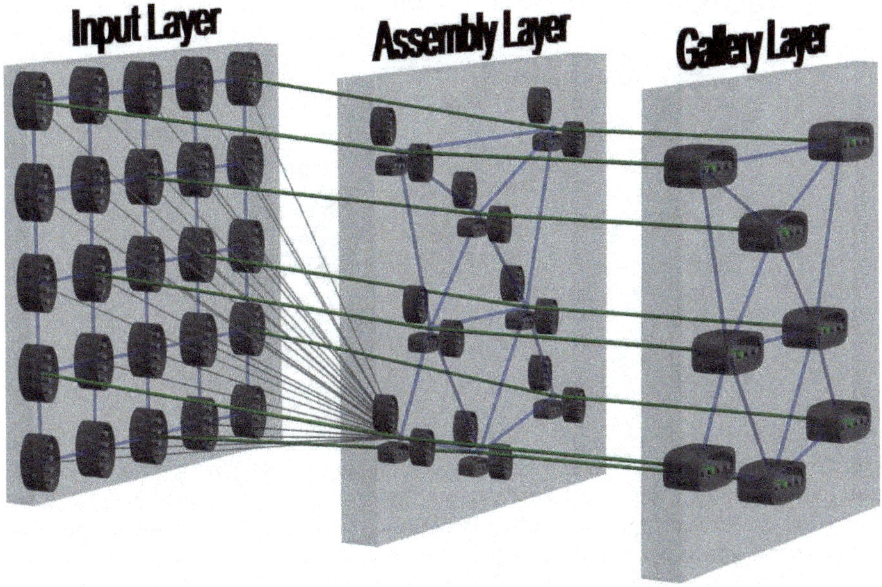

Fig. 3.7 Architecture of the network. The gray oval structures represent columns (the vertical ones feature columns, the horizontal ones decision columns), with units as lighter cylinders inside. The numbers of units and columns shown here are chosen exemplarily for visualization purposes only and are not identical to the real numbers of units used in by us. The Input Layer is organized in a rectangular grid (represented by the light lines connecting columns), while both the Assembly Layer and the Gallery Layer have a face graph topology. At each landmark in the Assembly Layer there are three columns, two feature columns of the Input Assembly and Gallery Assembly, and one control column. Input and Assembly are connected all-to-all (shown exemplarily for the left-lowermost point in the Assembly Layer), while Assembly landmarks are connected only to the same landmarks in the Gallery, but to all identity units there (see also Figure 3.8). The dark lines connecting the three layers and the subset of dark ($\widehat{=}$ activated) Gallery units represent a possible final state of the network.

use the standard Gabor transform, as described in Appendix A.2. As feature values we use the magnitude $\mathscr{J}$ of the responses, thus ignoring Gabor phase, to model complex cell responses (Hubel and Wiesel, 1977). Implicitly, Gabor phase is still represented by the positions of the feature columns in the input image. In applications using Gabor features it has turned out that with $K = 40$, as above, good results can be achieved (Wundrich et al., 2004). Performance increases for more wavelets, but 40 represents a good compromise between performance and computational cost.

Each Input Layer unit being responsive to a certain Gabor feature $\mathscr{J}_i^p$ at its position p on the input grid, the unit activities follow the dynamics (cf. (3.2)).

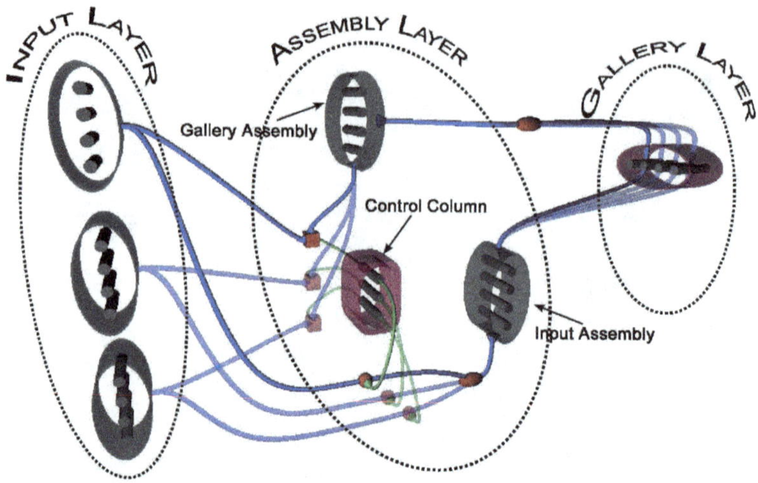

Fig. 3.8 Details of information processing by control units. Green lines represent signals to and from control units, blue lines show the pathways of visual signals between the layers of the system. Cube: Scalar product. Sphere: Modulation by control unit. Ellipsoid: Combination of several signals by averaging.

$$\tau \frac{d}{dt} x_i^{\mathcal{J}_P} = \mathcal{J}_i^P - x_i^{\mathcal{J}_P} \sum_{j=1}^{K} \mathcal{J}_j^P x_j^{\mathcal{J}_P}. \tag{3.5}$$

3.4.2 Assembly Layer

The Assembly layer integrates intermediate information from both the input image (represented in the *Input Assembly* units) and the gallery (represented by the *Gallery Assembly* units, see Figure 3.9). The role of the Input Assembly is to represent a normalized version of the input image, while the Gallery Assembly accommodates a weighted average of all Gallery faces. This information is organized in a face graph arrangement with $Q = 48$ landmarks (see Figure 3.6b).

Since the face graph in the Assembly Layer has to be able to represent many different faces, we determine its geometry by averaging over several hundred face graphs of individual faces: Assume M face graphs given, each with Q landmarks at positions. All coordinates are normalized relative to the full image width and height, going from 0 to 1. To discount possible different relative positions of the phase graphs in the coordinate system, the average graph is calculated in a translation-invariant fashion and centered around the image center. More specifically, the coordinates used from every single face graph to calculate the average graph are discounted by the amount that the center of mass of this face graph deviates from the image center $(0.5, 0.5)$. Let

$$\mathbf{c}_m = \frac{1}{Q} \sum_{q=1}^{Q} \mathbf{x}_{q,m}$$

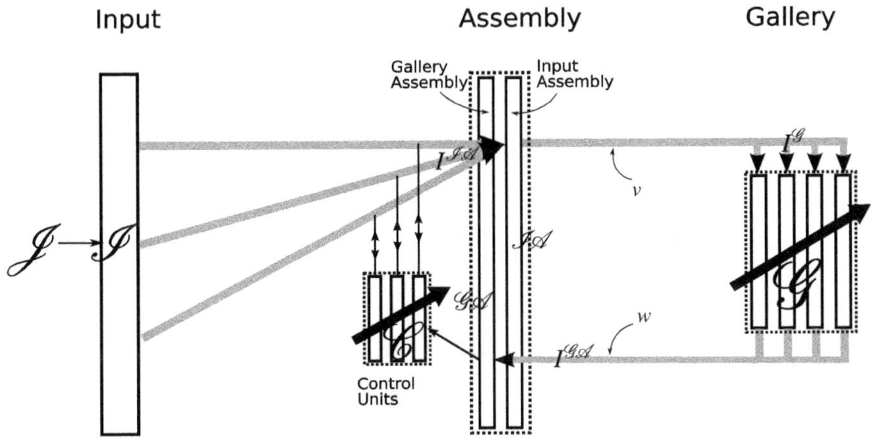

Fig. 3.9 Information flow in the network. Visual information in form of Gabor jets $\mathscr{J}$ extracted from an input image activates the Input Layer $\mathscr{I}$. It flows to the Assembly Layer (Input Assembly, $\mathscr{IA}$) and from there to the Gallery $\mathscr{G}$, where it activates via receptive fields v some memories more strongly than others. Information representing the active memories (stored in projection fields w analogous to v) flows back to the Gallery Assembly $\mathscr{GA}$. Information flow $I^{\mathscr{IA}}$ from the Input Layer to the Input Assembly is modulated by the control units $\mathscr{C}$, which in turn are driven by the similarity of those image patches in the Input Layer and the Gallery Assembly that they connect. By activating those control units that connect positions of the Input Layer containing similar information as the Gallery Assembly, the system effectively focusses on those parts of the input image that contain visual information most similar to the current reconstruction in the Gallery Assembly, formed by superposition of active units in the Gallery Layer. The thick black arrows represent the competition among the decision columns of which the Gallery and the control columns consist. The symbols correspond to those used in the text.

be the center of mass of a face graph m. Then we calculate the average position of a landmark q as

$$\mu_q = \frac{1}{M} \sum_{m=1}^{M} \mathbf{x}_{q,m} - \left(\mathbf{c}_m - \begin{pmatrix} 0.5 \\ 0.5 \end{pmatrix} \right),$$

which will produce an average graph with its center of mass at the image center. Additionally, it is possible to calculate the standard deviation of each average landmark position for a certain population of face images. This may be useful to assign different reliability to the different landmarks, or to define directions in which the average graph should be flexible. While we consequently calculated this information (see Figure 3.10), we did not use it in the scope of this work.

The columns of the Input Assembly and Gallery Assembly are feature columns, i.e. they integrate their inputs (defined below) according to (3.2). The input $I^{\mathscr{IA}}$ to the ith Input Assembly unit at position q of the face graph is a weighted sum of the ith Gabor feature at all grid positions p of the Input Layer, modulated by the respective control units:

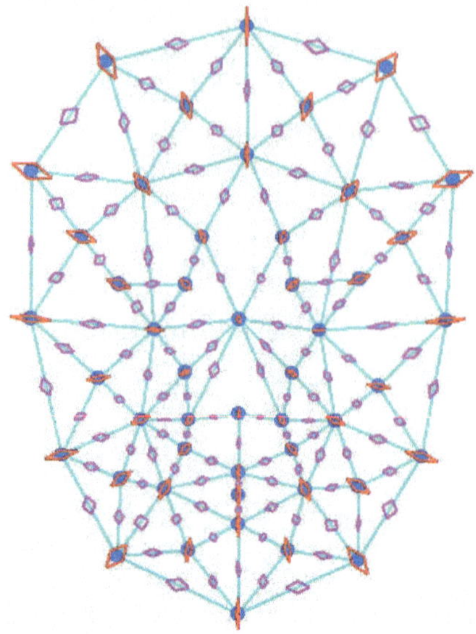

Fig. 3.10 Average face graph. An average face graph represents the geometry of a whole group of face images in a parsimonious yet meaningful way. Each node represents the mean position of a specific landmark in the whole group. The diamonds around the nodes denote the first and second moments of the standard deviation of landmark positions. The diamonds on the edges denote standard deviation of landmark distance. The graph shown here was constructed from 100 faces of the AR database (Martinez and Benavente, 1998).

$$I_{q,i}^{\mathscr{IA}} = \frac{1}{\sqrt{P}} \sum_{p=1}^{P} \mathscr{C}_{p,q} \mathscr{I}_{p,i}, \tag{3.6}$$

with $\mathscr{C}_{p,q}$ the output strength of the dynamic link (see below) controlling the flow of the output of Input column $\mathscr{I}_p$ to Input Assembly column $\mathscr{IA}_q$.

The input $I^{\mathscr{GA}}$ to a Gallery Assembly column at position q is the superposition of all Gallery projective fields $w_{q,m}$, weighted by the activity of the respective Gallery unit $\mathscr{G}_{q,m}$:

$$I_{q,i}^{\mathscr{GA}} = \frac{1}{\sqrt{M}} \sum_{m=1}^{M} w_{q,m,i} \mathscr{G}_{q,m}, \tag{3.7}$$

with the "efferent weight" $w_{q,m,i}$ representing the strength of Gabor feature i in landmark q of Gallery image m (of M in total).

Control units

The Assembly Layer also contains the *control units* mentioned above, which mediate the signal coming in from the Input Layer. These control units provide potential connections (*dynamic links*) between every Input Layer point to every point in the Input Assembly. The activity of the control units is driven by the feature similarity of the corresponding points in the Input Layer and the *Gallery* Assembly. That is, the similarity between the non-normalized input face in the Input Layer and the weighted average face in the Gallery Assembly controls via the control units how input information flows to the Input Assembly. In that sense the control units define a geometric mapping between Input and Assembly Layer. Additionally to the feature similarity input, control units get support from neighboring control units that represent similar mappings (see Figure 3.11 and paragraph below for details.).

The dynamic links are decision units, meaning that their dynamics follow (3.1). The input $I^{\mathscr{C}}$ to a dynamic link $\mathscr{C}_{p,q}$ connecting input position p and assembly position q is given by the scalar product between both column outputs plus a topological interaction term:

$$I^{\mathscr{C}}_{p,q} = \sum_{i=1}^{K} \mathscr{I}_{p,i}\, \mathscr{GA}_{q,i} + \frac{c_{\text{top},\mathscr{C}}}{|\text{neighbors}|} \sum_{\tilde{p},\tilde{q}} f_{\text{top}}(p,q,\tilde{p},\tilde{q}) \mathscr{C}_{\tilde{p},\tilde{q}}, \tag{3.8}$$

where $c_{\text{top},\mathscr{C}}$ defines the maximal strength of topological interaction between control units (see below), and $|\text{neighbors}|$ is the number of topological neighbors the control column has in the face graph.

Topological cooperation among control units

As mentioned before, there is topological cooperation among the control units of the Assembly Layer. The purpose of this cooperation is to establish a continuous mapping between the different geometries of the Input Layer and the Input Assembly. A given dynamic link connects a specific column A of the Input Layer with a column B of the Input Assembly. Due to the geometry of both layers, both columns represent distinct positions $\mathbf{z}_A$ and $\mathbf{z}_B$ in retinal coordinates and internal image representation space, respectively. Consequently, the dynamic link between them represents a certain geometric distance $\mathbf{d}_i = \mathbf{z}_B - \mathbf{z}_A$.

The idea is now to have topological connections in order to support parallel or near-parallel dynamic links. Therefore we define the strength of a topological connection between any two dynamic links i and j whose columns are neighbors in the face graph through a monotonically decreasing function of their non-parallelity/disparity:

$$f_{\text{top}}(i,j) = f(\|\mathbf{d_j} - \mathbf{d_i}\|_2). \tag{3.9}$$

Here we use a linearly decreasing thresholded function of the form

$$f(y) = \max\left(0, 1 - \frac{y}{\beta}\right). \tag{3.10}$$

Thus topological interaction is always positive and acts only between more or less (depending on β) parallel neighboring links. This principle is depicted in Figure 3.11. To obtain the topological interaction f_{top} in (3.8) between two control units $\mathscr{C}_{p,q}$ and $\mathscr{C}_{\tilde{p},\tilde{q}}$, we first calculate from the coordinates of the columns they control in the Input and the Assembly Layer the geometric distances $\mathbf{d}_{p,q}$ and $\mathbf{d}_{\tilde{p},\tilde{q}}$ represented by them. From these we calculate the disparity of the two control columns according to (3.9) and the topological interaction via (3.10).

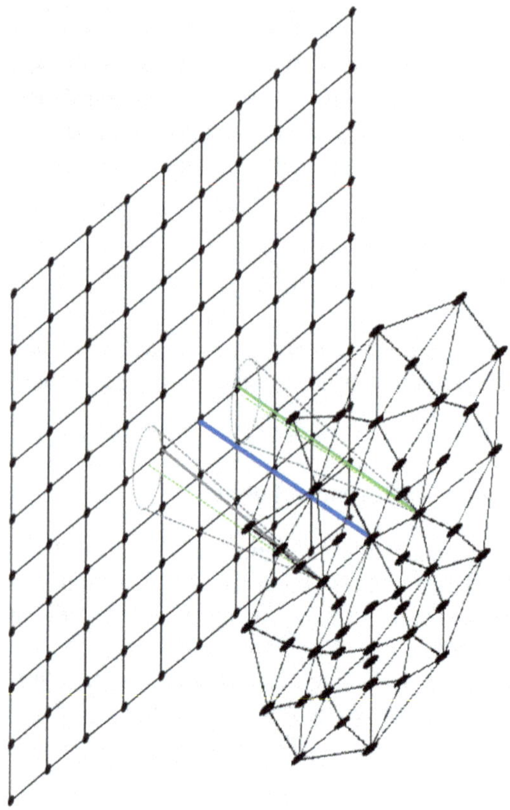

Fig. 3.11 Interaction among control units to achieve a topologically consistent (i.e., continuous) mapping. The unit controlling the blue link strengthens control units in neighboring columns that represent links of similar orientation. Maximal cooperation would occur with perfectly parallel links (the green dashed axes of the cones). Since in reality links in the network only exist to the nodes of the input grid (full lines), the strength of cooperation depends on the degree of parallelity with the dark link, equivalent to the distance of a link's end point from the cone center. The strength of cooperation with the blue link is visualized here by the shade of green of the respective link.

3.4.3 Gallery Layer

The Gallery Layer represents all M gallery face images in a face graph of Q decision columns. Each column corresponds to one landmark, with the units representing specific feature vectors for the individual faces at the respective landmarks by their afferent and efferent connections (see Figure 3.9). The units in the Input Assembly activate the Gallery units through receptive fields v representing the stored facial landmark features, activating more strongly units of faces that are similar to the normalized input image in the Input Assembly:

$$I^{\mathscr{G}}_{q,m} = \sum_{i=1}^{K} v_{q,m,i}\ \mathscr{IA}_{q,i} + \frac{c_{\mathrm{top},\mathscr{G}}}{Q} \sum_{\tilde{q}=1}^{Q} \mathscr{G}_{\tilde{q},m}. \tag{3.11}$$

Additionally, there is interaction among the Gallery units, with $c_{\mathrm{top},\mathscr{G}}$ defining how strongly Gallery units representing the same face cooperate. That is, all landmarks that belong to the same face cooperate, and at each landmark the corresponding features of all different faces compete.

The Gallery projects a weighted superposition of its stored faces to the Gallery Assembly through efferent weights w that are identical to its afferent weights v (cf. (3.7)). See Section 3.6 for some remarks on why we think this dual representation is crucial for any full visual system. Point-to-point comparison with the Input Assembly and competition among stored models leaves only the correctly recognized identity active in the end.

3.5 Results

We now simulate the dynamics defined in the above sections using natural images of faces as input and as memories in the gallery. The size of the network is determined by the following parameters. The input grid consists of $P = 400$ columns, all face graphs (Assembly and Gallery Layers) contain $Q = 48$ columns. Consequently, there is a total of $400 \times 48 = 19200$ control units. For representation of visual information in the Input and Assembly Layers, we use $K = 40$ Gabor wavelets. The number M of Gallery faces depends on the size of the database on which the system is tested. If not stated differently, we used the following settings and parameters. The radius for topological interaction among control units was $\beta = 0.05 \times$ imagesize. Maximum strength of this topological interaction was $c_{\mathrm{top},\mathscr{C}} = 3.5$. Cooperation strength between neighboring gallery units was $c_{\mathrm{top},\mathscr{G}} = 0.1$.

Note that we can numerically simulate dynamics (3.1) without specifying a value of the time constant τ. As long as the simulation time T remains constant relative to τ, simulation results will be independent of τ. The question of how the time-course of the dynamics translates to recognition times in biological terms does, however, crucially depend on the actual choice of τ. We chose a time-constant of $\tau = 0.2$ms and the length of the v cycle as $T = 400\,\tau$. This results in a system that selects the winning sub-populations of its decision columns in about 80ms (with $T = 400\,\tau$, compare Figure 3.4). The whole network could consequently converge to a face position and identity within about the same time. A biophysical source for this oscillating v signal might be oscillations of brain activity in the gamma or theta range (for the role of oscillations in the brain for perception, see, e.g., Singer, 2003). Numerical simulations of a single column with explicitly modeled spiking neurons suggest an even smaller time-constant (see Lücke and von der Malsburg (2004) or compare Muresan and Savin (2007) for population activation times on the order of 10ms which suggest similarly fast deactivation times). All units have a small, but non-zero initial activity $x(0) = 0.01$. To integrate the Equations (3.1), we simply use the Euler method but adapt its time step dynamically to the average change of activity in the network in order to keep the system stable. For this, we regulate global network activity by controlling signals like the average change of columnar activity. Within a predefined maximal range, integration speed is decreased when these signals are too high, and it is increased when they fall below a certain threshold. We believe that such homeostatic processes are key ingredients to building large artificial networks that are adaptive and robust at the same time.

3.5.1 General Network Behavior

The units in the Input Layer, which receive input directly from the incoming image (cf. (3.5)), quickly converge to a state where they represent the input image via the different Gabor feature values at all grid positions. This information flows to the Input Assembly modulated by the activities of the control units (3.6) which connect every point in the Input Layer with every point in the Input Assembly. Since initially all control units have equal activity, this leads to a superposition of image information from all Input Layer points at each Input Assembly location, resulting in a feature-less, more or less homogeneous image in the Input Assembly (first image in Figure 3.12). In the Gallery Layer, all faces are equally active initially. The Gallery Assembly, which receives input from all Gallery units (3.7), will therefore initially receive a superposition of all Gallery faces, resembling an "average face" (like the first image in Figure 3.13).

To each control unit in the Assembly Layer a unique pair of feature columns is assigned, one in the Input Layer and the other one in the Gallery Assembly. The control units are driven by the similarity (expressed in terms of the scalar product) of the information stored in their dedicated feature columns, see (3.8). Therefore control units that connect points of the average face with similar input points will become stronger, while control units representing irrelevant matches will be weakened. Over the process of recognition, the activity distribution of the control units becomes more and more sparse, until it finally represents a unique mapping between the Input Layer and the Assembly Layer (see left column in Figure 3.12). Since purely local similarity of images can be quite ambiguous, the additional topological interaction among the control units is necessary in this process to achieve a globally consistent match. As the information flow from the Input Layer to the Input Assembly is modulated by the control units, the image in the Input Assembly will start to develop from a gray nondescript superposition to a more and more clear

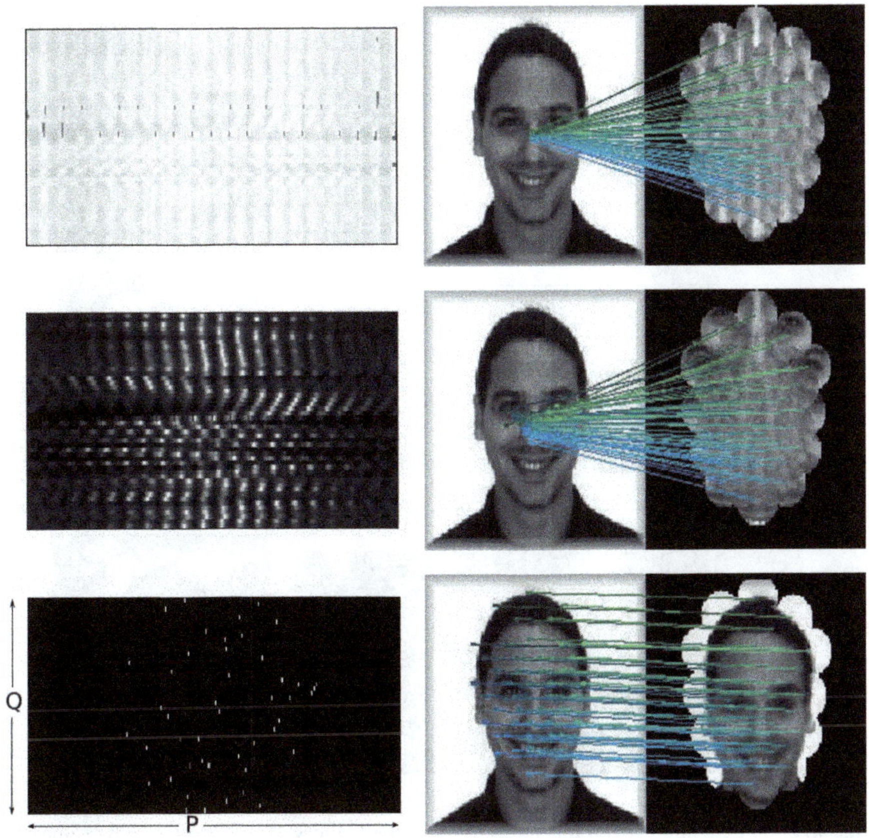

Fig. 3.12 The process (from top to bottom) of finding the correct mapping between the Input Layer and the Input Assembly. Each row shows the control unit activities on the left side (white means high relative activity; values are scaled individually for each image to exploit the full dynamic range.), and on the right first the constant input image, and then an image reconstructed from the activities of the 48 landmarks of the Input Assembly (each little patch is a superposition of the Gabor wavelets of all units of one column, weighted by the activity of the respective unit.). Initially, the control units all have nearly identical, but small activity, and therefore the Input Assembly receives a superposition of all input information, resulting in the same uniform image information at all landmarks (row one). With the control units developing a topologically consistent match between Input and Input Assembly (rows two and three), this image starts to differentiate towards a normalized (i.e. shifted and deformed if necessary) version of the input image. The mapping via the control units is also visualized by the colored lines connecting the input image with the Input Assembly. Each line represents the "center of mass" of a control column, i.e. the location in the input image where its units are pointing to as a group, weighted by their activity.

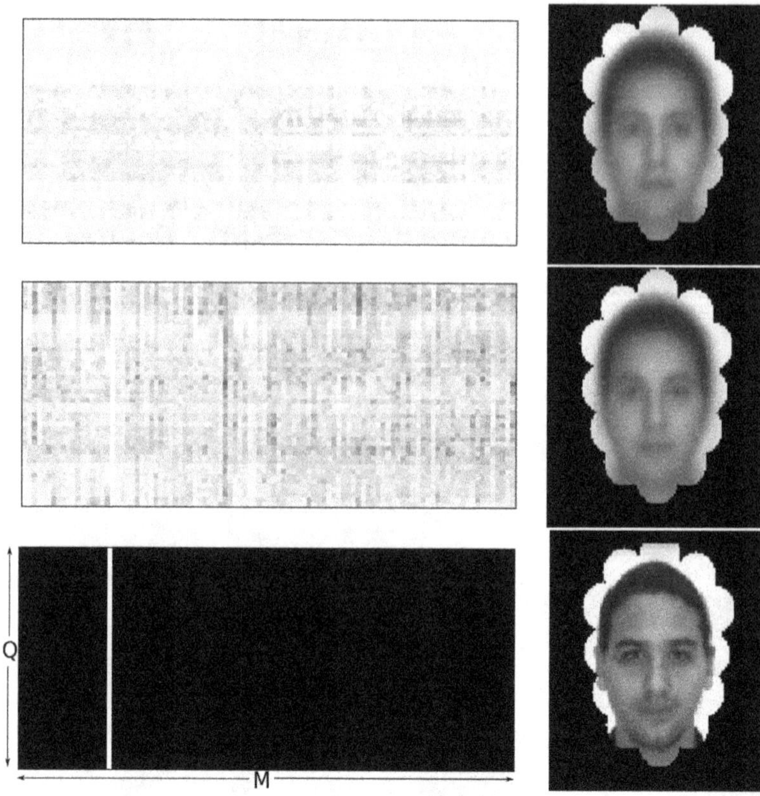

Fig. 3.13 Time course (from top to bottom) of the Gallery unit activities (left, values are scaled by the maximum activity for each image separately) and of the resulting image representation in the Gallery Assembly (right). The Gallery Assembly gets input from all Gallery units and thus contains an activity weighted average of all faces in the gallery. Initially, when all Gallery units are nearly equally active, this weighted average is a real average of all gallery faces, i.e. a mean face (uppermost row). With ongoing dynamics and rising competition, the Gallery units fitting the input image better get stronger, and the Gallery Assembly activity develops towards the respective gallery faces. Finally, only one unit of all Gallery columns is active, and the Gallery Assembly contains a representation of the image the system has recognized. Note that the input image presented to the system (shown in Figure 3.12) is not identical to the memory of the recognized face stored in the Gallery.

version of the input image (right column in Figure 3.12). It may be shifted and possibly distorted such that it conforms to the topology of the face graph of the Gallery Assembly.

The image information in the Input Assembly in turn acts as input to the Gallery units, where it gets filtered through the individual receptive fields of the units (3.11), exciting those units more that represent faces more similar to the input image. Owing to competition between the units of each Gallery column and cooperation among units of different landmarks representing the same face, the Gallery will start to favor some of the stored faces over others (cf. left column of Figure 3.13). This in turn changes the image in the Gallery Assembly from an average face to a superposition that is already biased towards one or several of the better fitting gallery faces (second and third face image of Figure 3.13). This sharpened target face now helps to position the normalized input image even more precisely, and so forth. In the final state, the Input Assembly will contain a shifted and maybe distorted version of the input image, while in the Gallery Layer the units of only one face are still active, and the Gallery Assembly contains a copy of that face of the Gallery that the system judges to be most similar to the input image.

In some cases, the system has not yet found a global optimum (i.e. a consistent match) after the first v cycle. In these cases, letting the system run for a second or third cycle usually produces the right result. In that sense, our system will always produce *some* result after a limited time, but it will continue to improve this result if given more time.

3.5.2 Position Invariance

Recognizing objects invariantly of their exact position is a crucial ability of visual animals. Position invariant recognition in humans is limited by the log-polar mapping from retina to cortex, which represents peripheral areas poorly, and is not perfect at the fovea either (see, e.g., Cox et al., 2005). As a general principle, however, position invariance is well accepted and forms the basis of psychophysical paradigms like "popout" (Humphreys and Heinke, 1998; Thornton and Gilden, 2007). Even processing of totally novel objects seems to be position invariant (Bar and Biederman, 1999; Fiser and Biederman, 2001), suggesting generic, object independent mechanisms for position invariant recognition.

The system presented in this book possesses such inherent position invariance. As was outlined in Section 3.4.2, control columns are not biased towards a certain *absolute* position in the input image, but only try to represent a similar *relative* shift as their neighbors. In that sense, the result of the matching process can be a match to any position of the input image as long as it is globally consistent.

To assess the position invariant matching capabilities of our system, we store a single face in the Gallery as a model image. For this experiment, we organize all layers as rectangular grids and use an Input Layer four times as large as the Assembly Layer. Thus the task of the system is now to detect somewhere in the large Input Layer that sub-image that resembles most closely the model face in the Gallery. Figure 3.14 illustrates how the model handles this. Of course, the system is also position invariant when doing recognition from galleries containing many models. For this we refer to the experiments in Section 3.5.4 (where the topology of the Assembly and the Gallery is again a face graph like in the standard system architecture).

Note that if the target face in an input image sits so far at the periphery that parts of it are not represented by the Input Layer anymore, the graph in the Assembly Layer cannot find complete correspondence in the Input Layer. in this case it is necessary to extend the topologic interactions of the control units to wrap-around

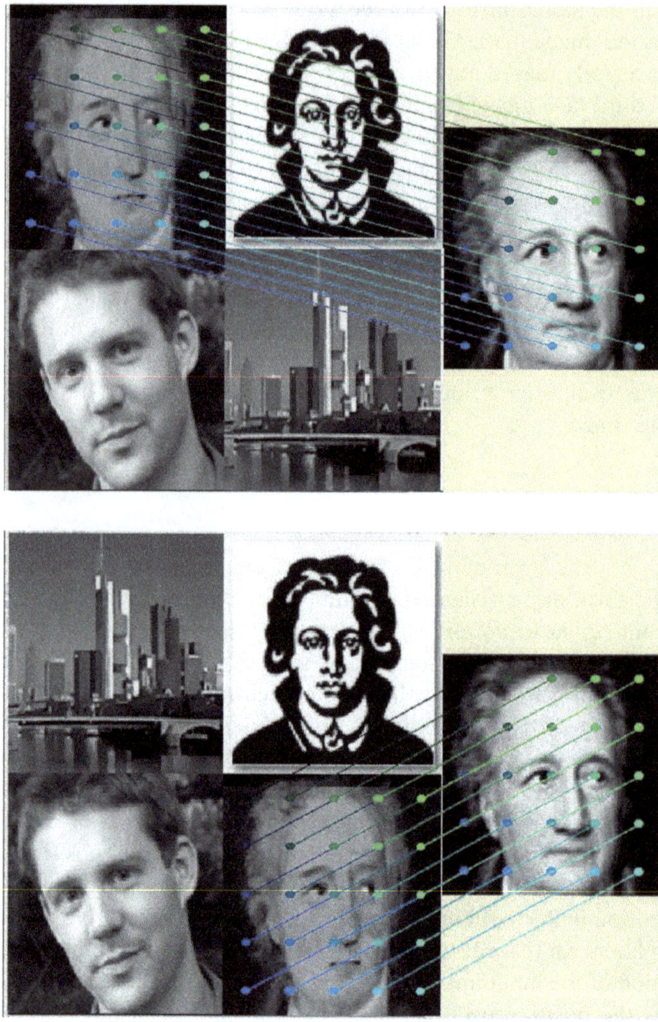

Fig. 3.14 Position invariance of our matching process. The Input Layer, containing a collage of several images, is four times as large as the Assembly Layer. Above, the final states of two matching processes are depicted. The input image is shown on the left and the model image to the right of it. Active dynamic links are shown as colored lines connecting input and model image. In both cases, the system successfully selects that sub-image in the Input Layer that resembles the model image most closely. Note that this process may take two ν cycles until the correct match is achieved, with the system sometimes first only reaching an inconsistent match due to interfering lateral influence from the other parts of the input image.

cooperation, enabling the nodes of the Assembly graph that would otherwise not find a partner to match to points on the opposite side of the input image so that the consistency of the graph matching is preserved.

3.5.3 Tests on Standard Databases

To quantitatively compare our system to other approaches, we tested it on the FERET (Phillips et al., 1998) and the AR (Martinez and Benavente, 1998) benchmark databases. We followed the testing protocols of (Phillips et al., 2000, the official FERET evaluation) and (Tan et al., 2005). The FERET database contains images of 1196 individuals (a sample of 30 of them is shown in Figure 3.15), while the subsets of the AR database used in (Tan et al., 2005) and by us contain 100 faces. Accordingly, the size of the Gallery when testing these databases is $M_{\mathrm{FERET}} = 1196$ faces and $M_{\mathrm{AR}} = 100$ faces, respectively.

In order to test the performance of a face recognition system, it is confronted with a gallery of images of all faces in the database, and is then asked to identify a different set of images containing pictures of (possibly a subset of) the faces in the gallery photographed under different conditions (cf. Figure 3.16). Often not only the

Fig. 3.15 A sample of 30 faces from the FERET database (Phillips et al., 1998)

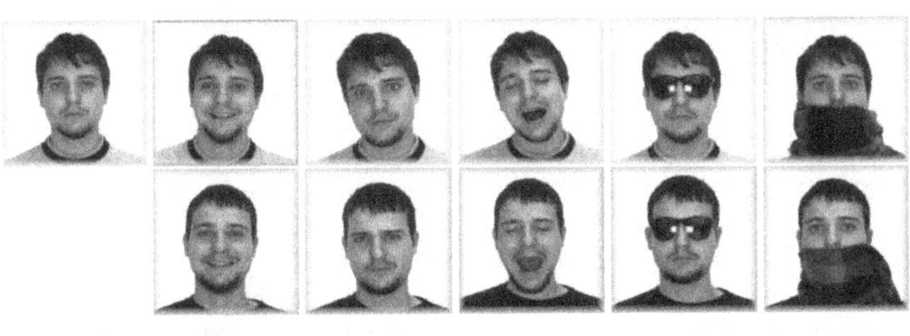

Fig. 3.16 Different expressions/conditions of the images used from the AR database. First row shows the gallery image, and subsequently the conditions smiling, angry, shouting, occlusion by sun glasses, occlusion with a scarf. The second row shows the same set of conditions taken at a later session on another day.

best match chosen by the system is recorded, but also the follow-up matches. This allows to construct *cumulative match scores*, the match score of rank n representing the fraction of test images whose correct match appears among the n best matches found by the system.

From the FERET database we used the following testing subsets: The set `fafb` contains photographs of 1195 individuals taken on the same day as the gallery images, but with the subjects showing a different facial expression. The set `Duplicate I` contains 722 images that were taken at least one day but less than 18 months after the gallery images. Finally, the set `Duplicate II` contains 234 images taken more than 18 months after the gallery images. The cumulative match scores of the system for this database are shown in Figure 3.17. The AR Face Database contains several testing subsets with images of the same 100 subjects that make up the gallery. The subsets b, c, and d contain images of the subjects smiling, expression anger, or screaming, respectively (see Figure 3.16). While those subsets were taken on the same day as the gallery images, the subsets h, i, and j show the subjects at a later session expressing those same three emotions. Subsets e and f show subjects wearing sunglasses and scarfs, and subsets k and l show the same situation at a later date. Cumulative match scores for this database are shown in Figure 3.18.

Table 3.1 Rank 1 match scores (in %) of the system, compared to those reported in the literature. The middle column shows the scores for the best performing system Wiskott et al. (1997) of the official FERET evaluation, and in brackets the average score of all 13 systems evaluated. The next column shows the performance of the two systems (SOM-Face/LocPb) proposed in Tan et al. (2005). The probe sets from the FERET database are the same as those of Figure 3.17, while for the AR database, the three emotion sets (b,c,d and h,i,j, respectively) and the two types of occlusion (e,f and k,l) have been averaged.

recognition rates [%]		our system	Phillips et al. (2000)	Tan et al. (2005)
FERET	fafb	95	95 (85)	92/-
	duplicate I	47	59 (40)	
	duplicate II	26	52 (22)	
AR	Emotion	91		95/82
	Em. duplicate	61		81/82
	Occlusion	73		96/81
	Occ. duplicate	36		56/51

Table 3.1 shows the performance of the system considering only the first match (i.e. cumulative match score for rank 1), and compares it to the recognition rates of the systems evaluated in (Phillips et al., 2000), and to the performance of (Tan et al., 2005). We can see that our system outperforms the average of the systems tested in the FERET evaluation, but does not reach the performance of the winner of this evaluation. Similarly, performance on the AR database is poorer than that of the better approach proposed in (Tan et al., 2005).

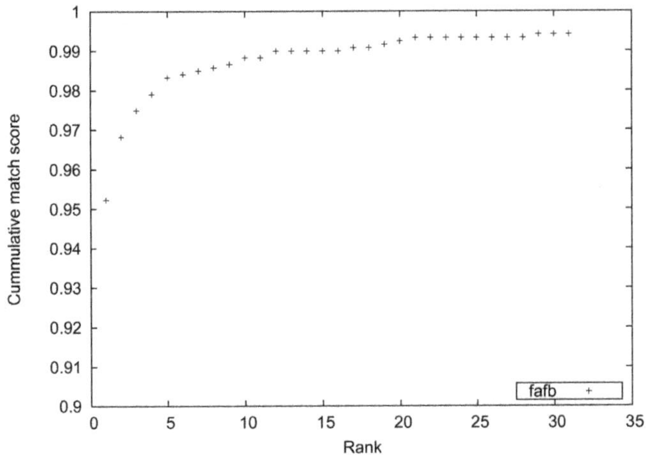

(a)

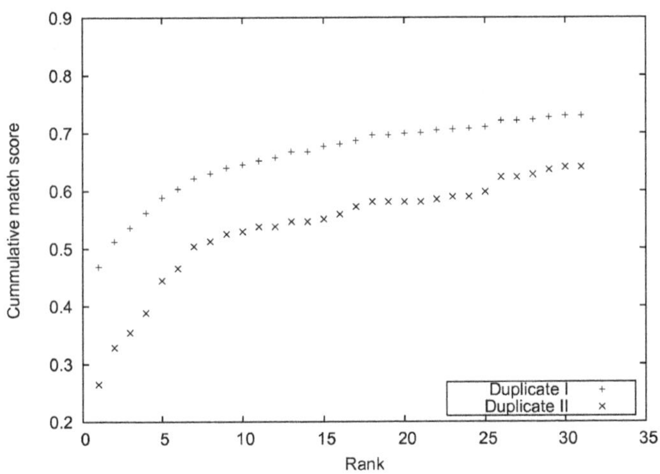

(b)

Fig. 3.17 Cumulative match scores for the FERET database. **(a)** Performance on the fafb dataset (different pictures taken in the same day). **(b)** Performance on the duplicate datasets.

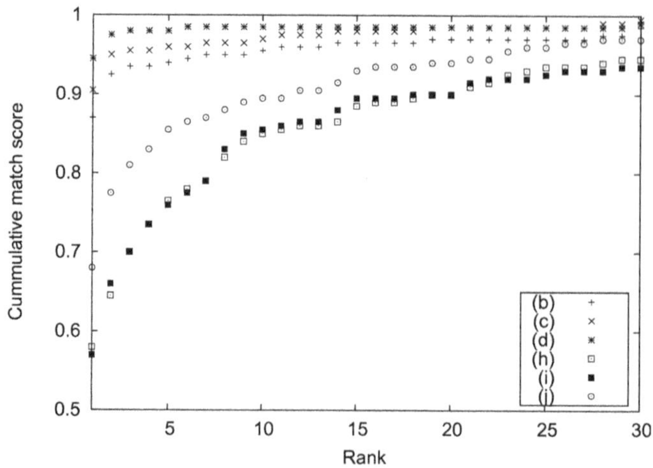

(a)

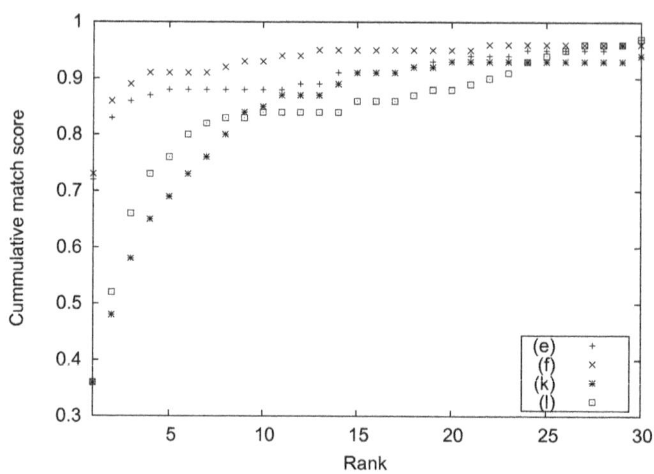

(b)

Fig. 3.18 Cumulative match scores for the AR database. **(a)** Performance on the emotion datasets. **(b)** Performance on the occlusion datasets.

We can conclude that while our system is definitely competitive, its performance does not reach that of top-notch, purely functionally motivated face recognition systems. However, we did not apply any parameter tuning to the system as tested here. For example, it turns out that performance of the model grows monotonously with input grid size, with our resolution of 20x20 points still being far from saturation. In fact, the winner of the FERET evaluation (Wiskott et al., 1997) uses Gabor wavelets from every pixel of the input image! Other parameters that could be optimized include the relative contribution of different landmarks, or the strength of topological interaction among the control units.

3.5.4 Attention Experiments

Attention and priming have a great influence on the way visual information is processed. Vision in animals and humans is not a static, reproducible process, but it depends on the current state of the organism. While this causes frustration with physiologists and experimental psychologists—who have great trouble constraining laboratory settings enough to reproduce their results—it is the basis of flexible, situation-specific behavior. For example, if we are looking for or expecting to see a certain object, we react faster and more accurately when the object finally comes into sight than if we were just watching passively (Duncan, 1984). This is called *semantic priming* or *object-based attention*. Similarly, when we focus our attention to a certain area in the visual field, information from this region is processed preferentially. This is called *spatial attention*. Some cases of visual information processing even require attention to work at all, for example serial search (for differences between the serial and parallel modes of visual search, see, e.g., Nakayama and Silverman, 1986 or Treisman and Sato, 1990) or detection of even large changes in images (*change blindness*, see, e.g. Simons and Rensink, 2005).

Both spatial attention and object-based attention effects can be realized in our system. Preactivating a subset of the control units at the beginning of a ν cycle (equivalent to spatial attention) results in a bias in favor of a specific location. For a large input image containing several faces, the system then preferentially processes and recognizes objects at that position (see Figure 3.19). Note that this way of preactivating pathways instead of feature activity merely biases the system towards a certain decision without distorting the content that is being processed. In comparison, this must be expected in the model of Deco and Rolls (2004), which implements spatial attention by increasing activity at the input level. Experimental findings (Luck et al., 1997), however, suggest that attention shifts effective receptive fields (that is, pathways) without changing activity within the afferent layer. This is clearly in favor of a correspondence based explanation of spatial priming.

Similarly, the priming of objects is possible with the help of preactivating search images in the form of arbitrary combinations of facial features in the Gallery Layer, leading to preferential detection and recognition of a similar face among several others in the visual field (see Figure 3.20). In distinction, preactivation of object-representing nodes in feature-based systems would not generate an explicit search image at lower layers but a large unstructured activation of *all* features that

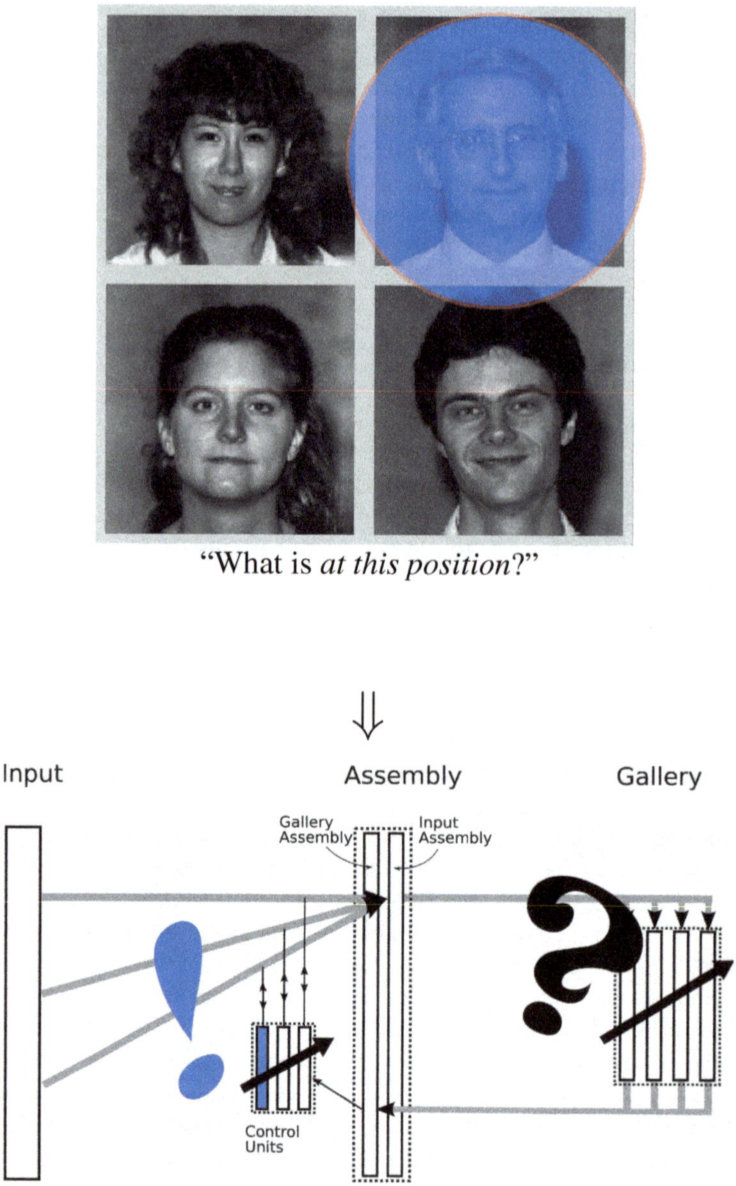

Fig. 3.19 Spatial attention. The system's attention is guided to a certain position in the large input image (visualized by the circle around one of the faces) by preactivating the control units pointing to the respective area of the Input Layer. This very weak preactivation is enough to bias the control units into matching the Assembly Layer with the attended face, and consequently, the person behind that face is recognized. Applying no attention results in a competition between the different input faces that can take several v cycles, until the most salient face present in the Gallery is selected and recognized. In that sense, spatial attention speeds up processing.

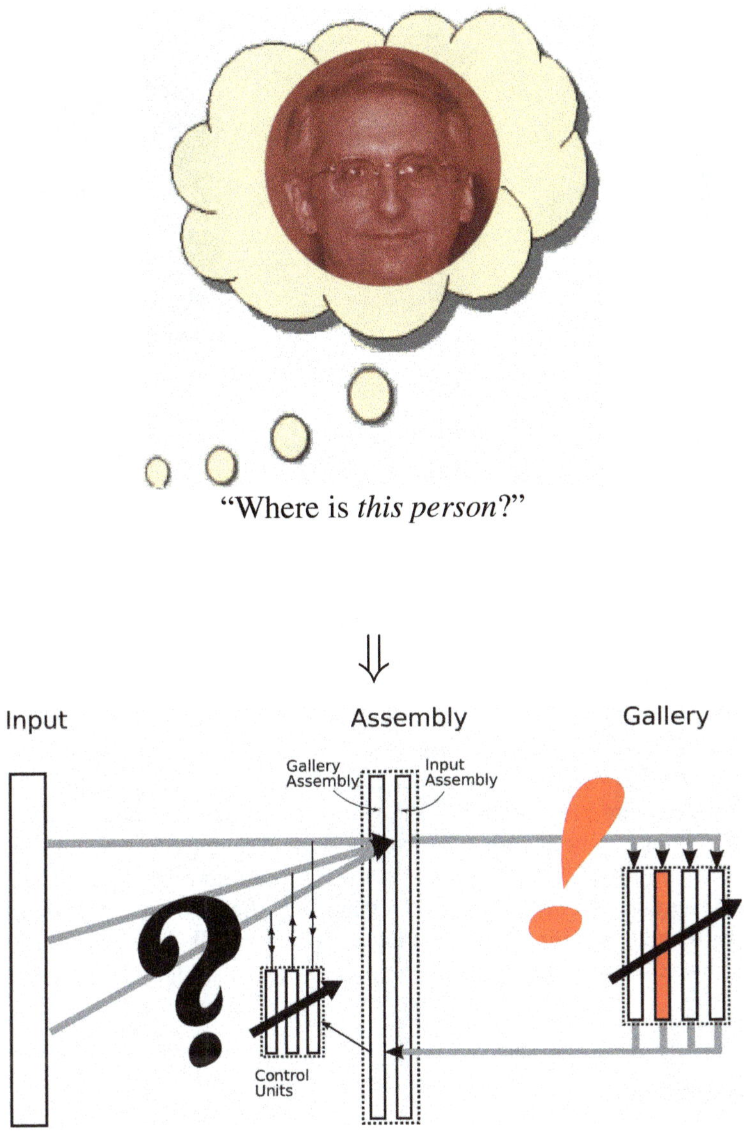

Fig. 3.20 Object search. The system can be primed for a specific face by slightly preactivating the respective Gallery units. This leads to the system searching the input image for a similar face and selecting and recognizing this face. If the face is not present in the input image, the system selects the most similar input object and recognizes it (which can result in a different set of Gallery units winning than originally preactivated).

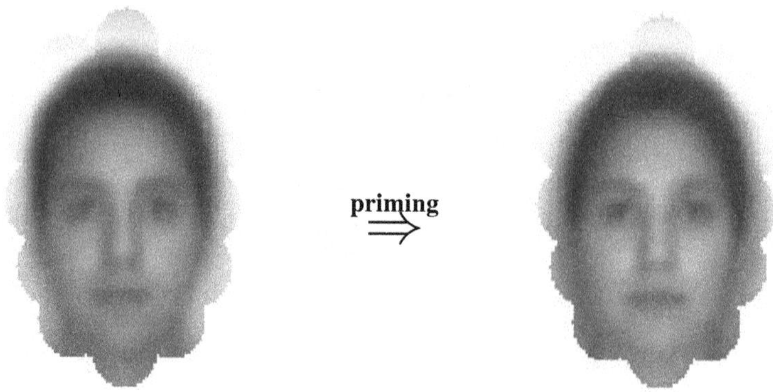

Fig. 3.21 Activity of the Gallery Assembly after priming of female faces. When the Gallery units representing female faces are preactivated, the visual information present in the Gallery Assembly changes from the initial neutral average face to an average female face, before then becoming more dedicated in the subsequent recognition process.

potentially could give rise to the primed object. It is quite unlikely that this would result in useful object priming, at least not in a field like face recognition, where the target object is not defined by a pure enumeration of its features, but rather by its specific arrangement.

Instead of priming a specific face, it is also possible to bias a large group of faces representing a certain type of face, for example a "female" face. When all faces of women are preactivated in the Gallery, the initial activity in the Gallery Assembly changes (see Figure 3.21 from a neutral average face to a prototypical female face (prototypical for the database being used, that is). Figure 3.22 shows how the system, having been primed on female faces, has chosen and recognized one of the two women present in the input image (which one it chooses is not specified by us in this case, so it chooses the one that has more similarity with the average female face).

3.6 Discussion

The model presented in this chapter is fully neural and combines findings from psychophysics, imaging studies, and physiology, while still performing competitively on benchmark tests for face recognition. The basic building block of the system is a model of the cortical column that makes use of a for stimulus representation, thus allowing significantly faster computations than with rate codes of single neurons (see, e.g., van Vreeswijk and Sompolinsky, 1998). An essential ingredient of the model are dynamic links, synaptic connections that are modulated by the activity of control units, whose activity in turn is controlled by signal comparisons. There is strong experimental evidence that receptive fields of neurons are not static (see (Luck et al., 1997) and other references cited in Chapter 2), suggesting the ex-

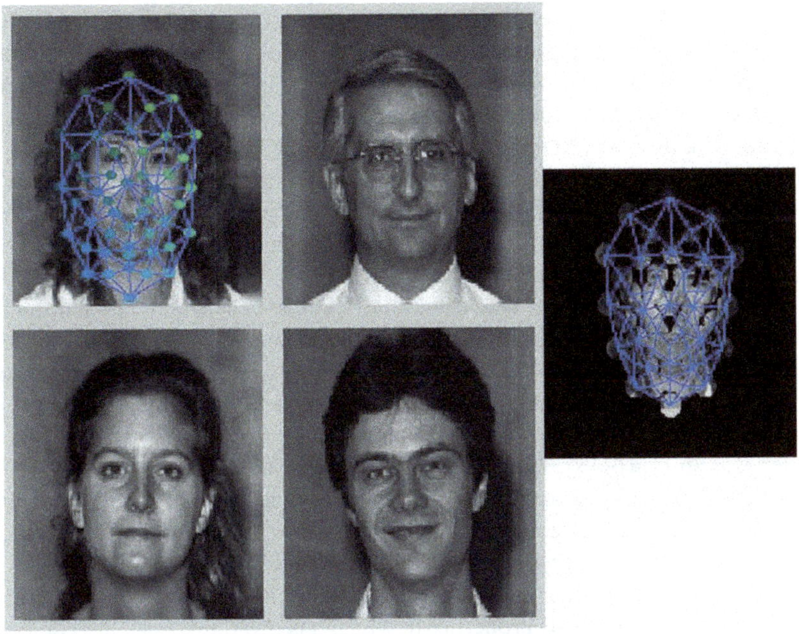

Fig. 3.22 Final result of a priming experiment. The Gallery units representing female faces were preactivated. This biases the system towards detecting female faces. The matching process locks onto that part of the input image that resembles most the female average face represented by the Gallery Assembly, and subsequently recognizes this person. The face graph landmarks in the input image (left) are positioned at the population averages of the active control units, indicating the correspondence the system has found. The right image shows the information present in the Input Assembly.

istence of dynamic links. For a discussion of further evidence for dynamic routing of information in the brain, see Section 4.2. Other models in the literature argue for similar concepts, like the control units of (Olshausen et al., 1993) or Sigma-Pi neurons (Weber and Wermter, 2007). For possible physiological mechanisms of modulation of neural signals, refer to Section 4.4.2.

In anatomical terms, the different layers of our model can be interpreted as follows. The input layer represents incoming image information by Gabor wavelets, which resemble the receptive field properties in primary visual cortex (V1). The biological counterpart of the Assembly Layer would be an area like central or anterior inferotemporal cortex. Neurons here respond to stimuli from large parts of the visual field, and they code for complex shapes similar to the face parts represented by the Assembly Layer (Tanaka, 1996, 2003). The fact that information about object position and scale can be read out from IT neurons (Hung et al., 2005), which disagrees with the assumptions made by pure pooling-models, points to the possibility of our control units residing there as well. Of course, in the cortex the mapping from V1 to IT does not happen directly, but via intermediate stages like

V2 and V4. This is not accounted for in our current model, but can be included in future extensions, as described in Chapter 4. Finally, the Gallery of the model might correspond to an area like the (FFA), which is specialized for face recognition (Kanwisher and Yovel, 2006; Tsao et al., 2006). Note that the decision *that* something is a face is not modeled by us. Also in the brain, this appears to happen outside of FFA. Summerfield et al. (2006) find neurons in medial frontal cortex that are selectively active when subjects have to make a face vs. non-face decision, independently of face *identification*. Likewise, prosopagnosia patients recognize objects as faces but cannot identify them (Zhao et al., 2003). As discussed before, face recognition is special because faces have a generic shape, but recognition from thousands of individuals requires high sensitivity to detailed differences. This might become possible through competitive interaction, which in fact is the mechanism by which recognition happens in our Gallery Layer, in the small and compact FFA (Kanwisher, 2006). Apart from faces, there is evidence suggesting that FFA can also serve as an area of expertise (Tarr and Gauthier, 2000; Gauthier et al., 2000) for other object classes. In the same sense, our model is not confined to face recognition, but could be used for recognition of any kind of object type that has a prototypical shape and requires high sensitivity to small differences among objects.

Our system is generative in the sense that it explicitly represents both the recognized object and the extrinsic properties to which it is invariant (especially object position). This has several consequences. First, the model requires top-down weights to re-generate the input, and these weights are identical to the bottom-up weights through which the Gallery units are activated. This existence of (not necessarily identical) weights in both directions is typical for generative models. The influential model of Olshausen and Field (1997) for learning V1 receptive fields makes use of both "analysis" and "synthesis functions", the Helmholtz machine (Dayan et al., 1995) has distinct "recognition" and "generative weights", and also current models for object recognition like (Murray and Kreutz-Delgado, 2007) use explicit bottom-up and top-down weights. It is hard to imagine a generative model without this dual representation. Let me note that this aspect of our model reflects anatomical reality, with cortical top-down connections being at least as strong if not stronger than feedforward connections. The top-down pathways that are usually used to reconstruct the current percept can also send down information coding for an *expected* percept. In that way, the system can be prepared for certain objects or locations, replicating typical attentional and priming effects. We explored these capabilities in Section 3.5.4.

Finally, object representations in generative models are *explicit*. A cardinal cell of a typical feature-based model represents objects or properties implicitly, as the activity of the cell does not express the structure of the object, which is only implicit in the synaptic patterns that define its firing condition. An explicit representation, on the other hand, cannot only detect presence of the object, but can also reproduce the appearance of the object under many different conditions, and may, e.g., relay this rich information to recipients elsewhere in the brain. An explicit representation therefore forms not just a sample point in appearance space but represents a whole space of variations. However, in distinction to the general, all-purpose representation in the retina, the ultimate goal of the visual system is to lead to explicit representations that encode the meaningful parts of a scene, like object identities and their properties, in a combinatorial and invariant code that is useful for subsequent action planning or motor control. It is these explicit representations that give us the feeling of being in direct contact with the visual reality out there.

Where does our model stand in the light of these distinctions? The representation in the Input Layer is explicit and uncommitted, and its generality is restricted only by the incomplete sampling of Gabor space. The representation in individual units in the Gallery Layer is implicit, highly specific and completely committed (to individual landmarks in individual faces). Given its activity state, the Gallery Layer creates, via its output connections to the Gallery Assembly, an explicit representation of a face, which for low levels of the inhibition is still uncommitted to an individual (see Figure 3.13, upper-most image), and at the end of the selection process corresponds to an explicit representation fully committed to one individual face. While models of object recognition that explicitly reconstruct the input are common in probabilistic modeling (conceptually discussed e.g. in Yuille and Kersten, 2006), it is still unclear, how such reconstructions are realized by the brain. Hopefully this book can contribute to deepening our insight in this respect.

Surely, the model still leaves open a number of problems for future work. As is, the model is invariant only to translation and needs to be generalized to changing scale and orientation (which will require dynamic relinking of feature connections, see Sato et al., 2007) as well as other image transformations such as changing illumination and perspective deformation. Another important extension of the system is autonomous learning of the contents of the Gallery.

An unrealistic aspect of the model as presented here are the direct dynamic links from all positions in the Input Layer to all positions of the Input Assembly. This would require an unrealistic number of fibers converging on any target unit. Therefore, we have investigated how the fiber convergence numbers and the overall number of neural elements required for correspondence finding can be reduced by distributing the process over an architecture of several stages. We turn to this topic in the next chapter.

References

Adler, A., Schuckers, M.E.: Comparing human and automatic face recognition performance. IEEE Trans. Syst. Man Cybern B Cybern. 37(5), 1248–1255 (2007)

Amaral, D.G., Schumann, C.M., Nordahl, C.W.: Neuroanatomy of autism. Trends Neurosci. 31, 137–145 (2008)

Bar, M., Biederman, I.: Localizing the cortical region mediating visual awareness of object identity. In: PNAS, vol. 96, pp. 1790–1799 (1999)

Biederman, I.: Recognition-by-components: a theory of human image understanding. Psychol. Rev. 94(2), 115–147 (1987)

Biederman, I., Kalocsai, P.: Neurocomputational bases of object and face recognition. Phil. Trans. Roy. Soc. B 352, 1203–1219 (1997)

Bienenstock, E., von der Malsburg, C.: A neural network for invariant pattern recognition. Europhysics Letters 4(1), 121–126 (1987)

Buxhoeveden, D.P., Casanova, M.F.: The minicolumn hypothesis in neuroscience. Brain 125, 935–951 (2002)

Cox, D., Meier, P., Oertelt, N., DiCarlo, J.J.: 'breaking' position-invariant object recognition. Nature Neuroscience 8(9), 1145–1147 (2005)

Dantzker, J.L., Callaway, E.M.: Laminar sources of synaptic input to cortical inhibitory interneurons and pyramidal neurons. Nature Neuroscience 3(7), 701–707 (2000)

Daugman, J.G.: Two-dimensional spectral analysis of cortical receptive field profiles. Vision Research 20, 847–856 (1980)

Dayan, P., Hinton, G.E., Neal, R.M., Zemel, R.S.: The helmholtz machine. Neural Computation 7(5), 889–904 (1995)

Deco, G., Rolls, E.T.: A neurodynamical cortical model of visual attention and invariant object recognition. Vision Research 44(6), 621–642 (2004)

DeFelipe, J., Hendry, M.C., Jones, E.G.: Synapses of double bouquet cells in monkey cerebral cortex. Brain Research 503, 49–54 (1989)

Douglas, R.J., Martin, K.A.: Neuronal circuits of the neocortex. Annual Review of Neuroscience 27, 419–451 (2004)

Douglas, R.J., Martin, K.A., Witteridge, D.: A canonical microcircuit for neocortex. Neural Computation 1, 480–488 (1989)

Duncan, J.: Selective attention and the organization of visual information. J Exp. Psychol. Gen. 113, 501–517 (1984)

Eigen, M.: Selforganization of matter and the evolution of biological macromolecules. Naturwissenschaften 58, 465–523 (1971)

Favorov, O.V., Diamond, M.: Demonstration of discrete place-defined columns, segregates, in cat SI. Journal of Comparative Neurology 298, 97–112 (1990)

Favorov, O.V., Kelly, D.G.: Minicolumnar organization within somatosensory cortical segregates II. Cerebral Cortex 4, 428–442 (1994)

Fiser, J., Biederman, I.: Invariance of long-term visual priming to scale, reflection, translation, and hemisphere. Vision Research 41, 221–234 (2001)

Gauthier, I., Skudlarski, P., Gore, J.C., Anderson, A.W.: Expertise for cars and birds recruits brain areas involved in face recognition. Nature Neuroscience 3(2), 191–197 (2000), http://dx.doi.org/10.1038/72140

Gerstner, W.: Population dynamics of spiking neurons: fast transients, asynchronous states, and locking. Neural Computation 12(1), 43–89 (2000)

Goldstein, A., Harmon, L., Lesk, A.: Identification of human faces. Proceedings of the IEEE 59, 748–760 (1971)

Hubel, D.H., Wiesel, T.N.: Functional architecture of macaque visual cortex. In: Proceedings of the Royal Society of London - B, vol. 198, pp. 1–59 (1977)

Humphreys, G., Heinke, D.: Spatial representation and selection in the brain: Neuropsychological and computational constraints. Visual cognition 5, 9–47 (1998)

Hung, C.P., Kreiman, G., Poggio, T., DiCarlo, J.J.: Fast readout of object identity from macaque inferior temporal cortex. Science 310(5749), 863–866 (2005), http://dx.doi.org/10.1126/science.1117593

Jones, E.G.: Microcolumns in the cerebral cortex. Proceedings of the National Academy of Sciences, USA 97, 5019–5021 (2000)

Jones, J., Palmer, L.: An evaluation of the two-dimensional gabor filter model of simple receptive fields in cat striate cortex. Journal of Neurophysiology 58, 1233–1258 (1987)

Kanwisher, N.: Neuroscience. what's in a face? Science 311(5761), 617–618 (2006), http://dx.doi.org/10.1126/science.1123983

Kanwisher, N., Yovel, G.: The fusiform face area: a cortical region specialized for the perception of faces. Phil. Trans. R. Soc. B 361, 2109–2128 (2006)

Kirkpatrick, S., Gelatt, C.D., Vecchi, M.P.: Optimization by simulated annealing. Science 220(4598), 671–680 (1983), citeseer.ist.psu.edu/kirkpatrick83optimization.html

Körner, E., Gewaltig, M.-O., Körner, U., Richter, A., Rodemann, T.: A model of computation in neocortical architecture. Neural Networks 12, 989–1005 (1999)

Lades, Vorbrüggen, Buhmann, Lange, von der Malsburg, Würtz and Konen Lades, M., Vorbrüggen, J., Buhmann, J., Lange, J., von der Malsburg, C., Würtz, R., Konen, W.: Distortion invariant object recognition in the dynamic link architecture. IEEE Transactions on computers 42, 300–311 (1993)

Luck, S.J., Chelazzi, L., Hillyard, S.A., Desimone, R.: Neural mechanisms of spatial selective attention in areas V1, V2, and V4 of macaque visual cortex. J Neurophysiol. 77(1), 24–42 (1997)

Lücke, J., Keck, C., von der Malsburg, C.: Rapid convergence to feature layer correspondences. Neural Computation 20(10), 2441–2463 (2008)

Lücke, J., von der Malsburg, C.: Rapid processing and unsupervised learning in a model of the cortical macrocolumn. Neural Computation 16, 501–533 (2004)

Martinez, A., Benavente, R.: The AR face database, Technical Report 24, CVC (1998)

Messer, K., Kittler, J., Sadeghi, M., Hamouz, M., Kostin, A., Cardinaux, F., Marcel, S., Bengio, S., Sanderson, C., Poh, N., Rodriguez, Y., Czyz, J., Vandendorpe, L., McCool, C., Lowther, S., Sridharan, S., Chandran, V., Palacios, R.P., Vidal, E., Bai, L., Shen, L., Wang, Y., Yueh-Hsuan, C., Hsien-Chang, L., Yi-Ping, H., Heinrichs, A., Müller, M., Tewes, A., von der Malsburg, C., Würtz, R., Wang, Z., Xue, F., Ma, Y., Yang, Q., Fang, C., Ding, X., Lucey, S., Goss, R., Schneiderman, H.: Face authentication test on the BANCA database. In: Proceedings of the International Conference on Pattern Recognition, Cambridge, vol. 4, pp. 523–532 (2004)

Mountcastle, V.B.: The columnar organization of the neocortex. Brain 120, 701–722 (1997)

Mountcastle, V.B.: Introduction (to a special issue on cortical columns). Cerebral Cortex 13, 2–4 (2003)

Muresan, R.C., Savin, C.: Resonance or integration? Self-sustained dynamics and excitability of neural microcircuits. Journal of Neurophysiology 97, 1911–1930 (2007)

Murray, J.F., Kreutz-Delgado, K.: Visual recognition and inference using dynamic overcomplete sparse learning. Neural Computation 19(9), 2301–2352 (2007), http://dx.doi.org/10.1162/neco.2007.19.9.2301

Nakayama, K., Silverman, G.H.: Serial and parallel processing of visual feature conjunctions. Nature 320(6059), 264–265 (1986), http://dx.doi.org/10.1038/320264a0

Olshausen, B.A., Anderson, C.H., van Essen, D.C.: A neurobiological model of visual attention and invariant pattern recognition based on dynamic routing of information. Journal of Neuroscience 13(11), 4700–4719 (1993)

Olshausen, B.A., Field, D.J.: Sparse coding with an overcomplete basis set: a strategy employed by v1? Vision Research 37, 3311–3325 (1997)

Peters, A., Cifuentes, J.M., Sethares, C.: The organization of pyramidal cells in area 18 of the rhesus monkey. Cerebral Cortex 7, 405–421 (1997)

Peters, A., Yilmaz, E.: Neuronal organization in area 17 of cat visual cortex. Cerebral Cortex 3, 49–68 (1993)

Phillips, P., Flynnand, P., Scruggs, T., Bowyer, K., Chang, J., Hoffman, K., Marques, J., Min, J., Worek, W.: Overview of the face recognition grand challenge. In: IEEE Conference on Computer Vision and Pattern Recognition, pp. 947–954 (2005)

Phillips, P.J., Wechsler, H., Huang, J., Rauss, P.J.: The FERET database and evaluation procedure for face recognition algorithms. Image and Vision Computing 16(5), 295–306 (1998)
Phillips, P., Moon, H., Rizvi, S., Rauss, P.: The FERET evaluation methodology for face-recognition algorithms. IEEE Transactions on Pattern Analysis and Machine Intelligence 22(10), 1090–1104 (2000)
Ringach, D.L.: Spatial structure and symmetry of simple-cell receptive fields in macaque primary visual cortex. Journal of Neurophysiology 88, 455–463 (2002)
Rockland, K.S., Ichinohe, N.: Some thoughts on cortical minicolumns. Experimental Brain Research 158, 265–277 (2004)
Sato, Y.D., Wolff, C., Wolfrum, P., von der Malsburg, C.: Dynamic link matching between feature columns for different scale and orientation. In: Ishikawa, M., Doya, K., Miyamoto, H., Yamakawa, T. (eds.) ICONIP 2007, Part I. LNCS, vol. 4984, pp. 385–394. Springer, Heidelberg (2008)
Simons, D., Rensink, R.: Change blindness: past, present, and future. Trends Cogn. Sci (Regul. Ed.) 9, 16–20 (2005)
Singer, W.: Synchronization, binding and expectancy. In: Arbib, M. (ed.) The handbook of brain theory and neural networks, pp. 1136–1143. MIT Press, Cambridge (2003)
Summerfield, C., Egner, T., Greene, M., Koechlin, E., Mangels, J., Hirsch, J.: Predictive codes for forthcoming perception in the frontal cortex. Science 314(5803), 1311–1314 (2006)
Tan, X., Chen, S., Zhou, Z.-H., Zhang, F.: Recognizing partially occluded, expression variant faces from single training image per person with SOM and soft kNN ensemble. IEEE Transactions on Neural Networks 16(4), 875–886 (2005)
Tanaka, K.: Inferotemporal cortex and object vision. Annu. Rev. Neurosci. 19, 109–139 (1996)
Tanaka, K.: Columns for complex visual object features in the inferotemporal cortex: clustering of cells with similar but slightly different stimulus selectivities. Cereb. Cortex 13(1), 90–99 (2003)
Tarr, M.J., Gauthier, I.: Ffa: a flexible fusiform area for subordinate-level visual processing automatized by expertise. Nature Neuroscience 3(8), 764–769 (2000), `http://dx.doi.org/10.1038/77666`
Thornton, T.L., Gilden, D.L.: Parallel and serial processes in visual search. Psychol. Rev. 114(1), 71–103 (2007)
Treisman, A., Sato, S.: Conjunction search revisited. J Exp. Psychol. Hum. Percept Perform 16(3), 459–478 (1990)
Troncoso, E., Muller, D., Korodi, K., Steimer, T., Welker, E., Kiss, J.Z.: Recovery of evoked potentials, metabolic activity and behavior in a mouse model of somatosensory cortex lesion: role of the neural cell adhesion molecule (ncam). Cereb Cortex 14(3), 332–341 (2004)
Tsao, D.Y., Freiwald, W.A., Tootell, R.B.H., Livingstone, M.S.: A cortical region consisting entirely of face-selective cells. Science 311, 670–674 (2006)
van Vreeswijk, C., Sompolinsky, H.: Chaotic balanced state in a model of cortical circuits. Neural Computation 10, 1321–1372 (1998)
Weber, C., Wermter, S.: A self-organizing map of sigma-pi units. Neurocomputing 70(13-15), 2552–2560 (2007)
Wilson and Cowan Wilson, H.R., Cowan, J.D.: A mathematical theory of the functional dynamics of cortical and thalamic nervous tissue. Kybernetik 13, 55–80 (1973)
Wiskott Wiskott, L.: The role of topographical constraints in face recognition. Pattern Recognition Letters 20(1), 89–96 (1999)

Wiskott, L., Fellous, J.-M., Krüger, N., von der Malsburg, C.: Face recognition by elastic bunch graph matching. IEEE Trans. on Pattern Analysis and Machine Intelligence 19(7), 775–779 (1997), http://www.cnl.salk.edu/ wiskott/Abstracts/WisFelKrue97a.html

Wiskott, L., von der Malsburg, C.: Face recognition by dynamic link matching. In: Sirosh, J., Miikkulainen, R., Choe, Y. (eds.) Lateral Interactions in the Cortex: Structure and Function, Austin, TX. The UTCS Neural Networks Research Group, vol. 11, Electronic book (1996), www.cs.utexas.edu/users/nn/web-pubs/htmlbook96/, http://www.cnl.salk.edu/~wiskott/Abstracts/WisMal96c.html ISBN 0-9647060-0-8

Wolfrum, P., Lücke, J., von der Malsburg, C.: Invariant face recognition in a network of cortical columns. Proc. International Conference on Computer Vision Theory and Applications 2, 38–45 (2008)

Wolfrum, P., von der Malsburg, C.: Attentional processes in correspondence-based object recognition. In: Proc. COSYNE, p. 330 (2008)

Wolfrum, P., Wolff, C., Lücke, J., von der Malsburg, C.: A recurrent dynamic model for correspondence-based face recognition. J. Vis. 8(7), 1–18 (2008), http://journalofvision.org/8/7/34/

Wundrich, I.J., von der Malsburg, C., Würtz, R.P.: Image representation by complex cell responses. Neural Computation 16(12), 2563–2575 (2004), http://dx.doi.org/10.1162/0899766042321760

Würtz, R.P.: Multilayer Dynamic Link Networks for Establishing Image Point Correspondences and Visual Object Recognition, Verlag Harri Deutsch, Thun, Frankfurt am Main (1995)

Yoshimura, Y., Dantzker, J.L.M., Callaway, E.M.: Excitatory cortical neurons form fine-scale functional networks. Nature 433(7028), 868–873 (2005)

Yuille, A., Kersten, D.: Vision as Bayesian inference: analysis by synthesis? Trends in Cognitive Sciences 10(7), 301–308 (2006)

Zhao, W., Chellappa, R., Phillips, P.J., Rosenfeld, A.: Face recognition: A literature survey. ACM Computing Surveys 53(4), 399–458 (2003)

Chapter 4
Switchyards—Routing Structures in the Brain

In the previous chapter, we discussed a mechanism for correspondence finding and recognition. For simplicity, we assumed a given, rather simple all-to-all connectivity for correspondence finding between Input and Assembly Layer. In this chapter, we will investigate in detail how such a connectivity structure should look like to be biologically more realistic.

We first introduce the idea of multistage routing (Section 4.1) and discuss physiological evidence for it (Section 4.2). We proceed to derive an optimized architecture for routing in Section 4.3. These optimal architectures were termed *Switchyards* (Wolfrum, 2008), referring to the networks of tracks and switches that are used to route and reassemble railway trains. In Section 4.4, the resulting networks are discussed in functional and physiological terms, before the chapter concludes with Section 4.5. Parts of the material presented in this chapter, including figures 4.1, 4.2, 4.3, 4.5, and 4.7 have been previously published in (Wolfrum and von der Malsburg, 2007).

4.1 Multistage Routing

As discussed in Chapter 2, an important capability of biological vision systems is invariant object recognition. The same object seen at different position, distance, or under rotation leads to entirely different retinal images which have to be perceived as the same object. Only one of these variances, translation, can be compensated by movements of the eye. The approximate log-polar transform which takes place in the mapping from retina to cortex (Schwartz, 1977), on the other hand, replaces some kinds of transformations (scale, rotation) by others (translation on the cortex) and therefore does not fully explain invariant recognition, either. Invariant recognition, and how the visual system performs it, remains a topic far from being understood.

Invariance does *not* mean insensitivity to the spatial arrangement of visual information. While object recognition in our brain is invariant with respect to the above-mentioned transformations, it is very sensitive to small differences in the retinal activity pattern arising from, say, seeing both of your twin sisters shortly after each

P. Wolfrum: Info. Routing, Corres. Finding, and Object Recognition in the Brain, SCI 316, pp. 69–89.
springerlink.com © Springer-Verlag Berlin Heidelberg 2010

other. We believe that the only way a brain can solve these two competing problems realistically is to have a general, object-independent mechanism that compensates variance transformations without distorting image information. Thus the image can be conveyed in a normalized frame of reference to higher brain areas for recognition. Such a mechanism requires a routing network providing physical connections between all locations in the visual input region (V1) to all points in the target area (like IT). In addition, neural machinery is required to control these connections.

The necessity of dynamic information routing was appreciated early on in vision research (Pitts and McCulloch, 1947). Especially the finding that in posterior parietal cortex a frame of reference transformation takes place from retinal coordinates in primary visual areas to head-centered coordinates in higher areas (e.g., cf. Duhamel et al., 1997) has received a lot of modeling attention (e.g. Pouget and Sejnowski, 1997; Zipser and Andersen, 1988). For object recognition, on the other hand, the idea of dynamic routing is much less common, maybe because here its importance is not as evident as for coordinate transforms. Nevertheless, there have been models for routing in object recognition. In a sense, correspondence-based models for recognition like the one discussed in Chapter 3 dynamically route visual information, alas in a very basic way, by connecting input and memory domains in an all-to-all manner. This, however, is biologically not very plausible. To allow correspondence finding at acceptable resolution in the central parts of the visual field alone would require each point of the input to be connected to $\approx 100\,000$ points in the memory domain. This contradicts the fact that the number of inputs to typical neurons is rather on the order of 1 000 (Cherniak, 1990). Therefore, architectures have been proposed that distribute the process of visual information routing over several stages. These include the famous Shifter Circuits of Olshausen et al. (1993) or the SCAN model (Postma et al., 1997).

4.2 Physiological Background of Dynamic Routing

In spite of their essential advantages, systems that employ dynamic information routing are not as widespread in the modeling community as traditional neural networks with static connections. One reason for this may be that they require certain functional elements that static neural nets do not have. These are

1. a gating mechanism that allows neural signals to be modulated by the activity of other neurons and
2. neural units that perform this task of modulating other activities.

Although maybe more challenging to implement technically, networks with modulatory interactions between neurons are vastly more powerful than traditional neural networks (Durbin and Rumelhart, 1989; Koch, 1999). Moreover, there are plausible physiological explanations of how this additional functionality may be realized in the brain, as we will discuss in the following.

A gating mechanism requires that two neurons can *multiplicatively* influence the activity of a target neuron. Or, in other words, the activity of one neuron *modulates* the information transmission from another neuron to a target.

Experimental evidence for such multiplicative or modulatory interactions between neurons abounds. Gabbiani et al. (2002), e.g., have verified that specific neurons of locusts perform a multiplication of two input signals. A very prominent example in higher animals are the dopaminergic modulations of cortical input to the striatum (e.g., Freund et al., 1985; Nicola et al., 2000). Functionally, multiplicative interaction can be achieved by the input function of the target neuron having the general form $\sum_i \sum_j w_{ij} x_i x_j$, x representing the activities of the afferent neurons. Due to the subsequent multiplication and summing of inputs, such neurons are called *Sigma-Pi* neurons (e.g., Durbin and Rumelhart, 1989). Physiologically, such multiplication of inputs can be implemented by shunting inhibition at the same synapse (Volgushev et al., 1996; Kubota et al., 2007) or by supra-linear interaction of apical and distal inputs (Larkum et al., 1999; Schaefer et al., 2003) to a neuron. Another possibility for multiplication of neural signals is a nonlinear *activation* function. Tal and Schwartz (1997) note the fact that LIF neurons transform synaptic inputs approximately logarithmically into firing rates. Adding the *outputs* of two LIF neurons therefore gives the logarithm of the products of their inputs. For a wider review of cerebral gating mechanisms, refer to, e.g., (Salinas and Sejnowski, 2001).

The other necessary ingredient for routing networks are units that control the information flow through the routing network. Similarly to the control units of Chapter 3, they have to compare visual signals in the input and the memory domains, as well as incorporate the decisions of neighboring control units. In a multistage routing network, they additionally have to factor in the activity at intermediate stages. These requirements are met by the pulvinar nucleus in the thalamus, which is reciprocally connected to all stages of the ventral stream. Lesion studies in the pulvinar (Desimone et al., 1990) suggest that it is involved in directing visual attention, which would be a typical task of control units. On a similar note, the importance of the dorsal stream for processing the position of objects in the visual field suggests that control units might be located somewhere in that area and interact with the ventral stream from there. However, another, and maybe simpler explanation is that control units reside in the ventral stream itself, mixed with those units that represent and process feature information. This would make sense since it makes integration of information from and feedback of routing commands to the ventral stream much easier. While this idea contradicts the textbook view that the dorsal stream takes care of object position and the ventral stream is only responsible for object identity, there is growing evidence that these roles are not as separate as previously thought. Hung et al. (2005) have found that object position and scale can be read out from neurons in inferotemporal cortex (a high area of the ventral stream), while Konen and Kastner (2008) report representations of object identity in the dorsal stream.

Evidence that points specifically to the existence of multistage routing circuits as defined below is that the tangential spread of the basal dendrites and the number of dendritic branches and spines of pyramidal neurons increase along the ventral pathway (Lund et al., 1993; Elston and Rosa, 1998, 2000). This fits well with architectures like those of Figure 4.1. For discussions of evidence for dynamic information routing in general, see Sects. 2.3 and 3.6.

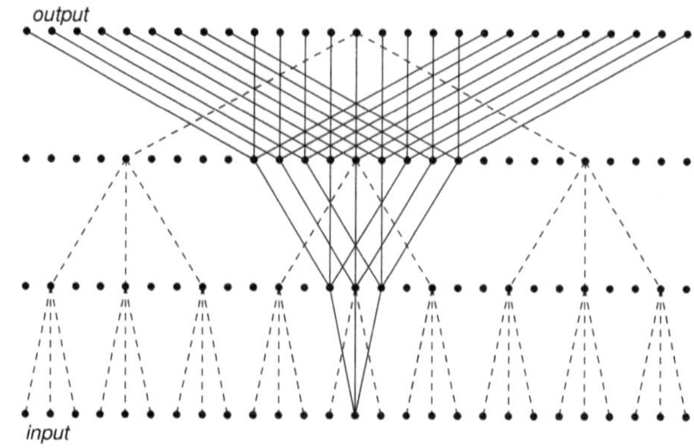

(a)

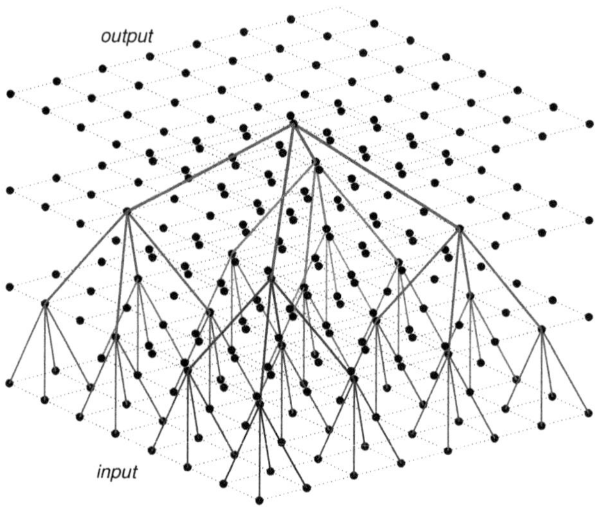

(b)

Fig. 4.1 Architectures for routing networks. (a) The one dimensional case, with $n = 27$ feature units and $k = 3$ routing steps, thus a fanout of $l = n^{\frac{1}{k}} = 3$. All feature nodes are shown (dots), but only selected links (lines), the others being shifted versions (with circular boundary conditions) of the shown links. All connections from one input node to the whole output stage are shown as solid lines, the connectivity between one output node and the whole input as dashed lines. (b) The two dimensional case, with $n = 64$, $k = 3$, and $l = 4$. Only the downward connections from a single output node are shown.

4.3 Optimized Architectures for Routing

What has been missing in previous models of multistage routing is the consideration of efficiency in terms of required neural resources. Different routing architectures require different numbers of intermediate feature-representing nodes (we will refer to them simply as *nodes*) and node-to-node connections (*links* from now on; if we mean both links and nodes, we will use the term *units*). In this chapter we will not discuss *how* connections in a routing circuit are controlled, we simply assume that the maintenance of a link and its control by a neural control unit have the same cost as feature nodes (for a deviation from this assumption, see the second part of Section 4.3.1). It is likely that cortical architectures have evolved which minimize this cost for the organism. Below, we therefore derive and analyze the routing network structure that minimizes the sum of all required units, both nodes and links. We call these optimal structures *Switchyards* in reference to the networks of tracks and switches that are used to route and reassemble railway trains.

In our analysis we will focus on two situations. In Section 4.3.1 we discuss routing between two cortical regions of identical size. This corresponds to perception of an already coarsely segmented object. In Section 4.3.2 we consider the architecture that must be present in real biological vision systems: Routing from a large input domain to a much smaller output domain, the first corresponding to the whole visual field, the second to a small higher-level target area engaged in object recognition.

4.3.1 Routing between Two Regions of the Same Size

Let us define a routing architecture with as few assumptions as possible:

- Input and output stages both consist of n image points. Each image point is represented by one feature unit (the extension of this to more than one feature per image point will be discussed below).
- The routing between input and output is established via $k-1$ intermediate layers of n feature units each. This number of layers is not predefined but will serve as the optimization parameter for minimizing required neural circuitry.
- Nodes of adjacent feature layers can be connected. For every such connection there exists one dynamic, neural unit that controls information flow in both directions. These units resemble the *control units* of (Olshausen et al., 1993). We assume here that one link (including its control unit) imposes the same "maintenance cost" as one feature node. If these costs are not identical, this can be accounted for with the parameter α introduced below.

Under these assumptions, what is the minimal architecture providing for each input node one separate pathway to every output node? For $k = 1$, the situation is clear: without any intermediate layer, every input node must be connected to all n output nodes. With intermediate layers, however, we can make use of a combinatorial code to achieve full connectivity, similar to "butterfly" computations used in the fast Fourier transform (Cooley and Connor, 1965): we assume that each input unit is only connected to l nodes of the adjacent intermediate layer (see solid lines in

Figure 4.1a). Each of these l nodes has in turn connections to l nodes of the following layer, and so on, until the output stage is reached. This method yields for every input node l^k pathways to the output stage, which are unique and lead all to different output nodes *if* we make sure that no two separate pathways merge again on the way to the output stage. An anatomically plausible way to meet this functional requirement is to let the spacing between target points increase exactly by the factor l from one link layer to the next, as shown in Figure 4.1a. The two-dimensional case is analogous, except that here the groups of nodes projecting to the same target are two-dimensional patches of l units with adequate spacing in between (see Figure 4.1b).

The connectivity described here agrees with the anatomical finding that loosely speaking, the spread of neuronal connections increases along the visual hierarchy. Perkel et al. (1986), for example, found a higher divergence of direct projections between V1 and V4 than between V1 and V2 or V3, respectively. In (Tanigawa et al., 2005), a four times larger spread of horizontal axons in inferotemporal cortex than in V1 was reported. Note, however, that the specific connectivity of the routing network is irrelevant for the results derived in the following. The only requirement is that the pathways of every input node be unique and lead to different output nodes.

In order to reach the whole output stage with these pathways, their number must equal the number of output nodes:

$$n = l^k.$$

From this we get the necessary neuronal fan-out at each stage as

$$l = n^{\frac{1}{k}}. \tag{4.1}$$

Let us now calculate how many nodes are needed to realize the routing architecture. Having $k-1$ intermediate layers means that a total of $(k+1)n$ feature nodes is required. All of these nodes, except those of the output layer, have l links to the next stage, resulting in a total of $knl = kn^{\frac{k+1}{k}}$ links. So the total number of units as a function of k and n is

$$N(k,n) = (k+1)n + kn^{\frac{k+1}{k}}. \tag{4.2}$$

As we can see in Figure 4.2, this number changes drastically with the number of intermediate layers being used. A direct all-to-all connectivity without any intermediate layers ($k=1$) is most expensive because the number of required links scales quadratically with n in this case. For a very large number of intermediate layers, on the other hand, the decrease in the required fan-out l is outweighed by the linear increase in nodes caused by additional layers. As we can see, there is a unique value k_{opt} for which the number of required units attains a minimum. To determine k_{opt}, we calculate the derivative of N with respect to k and set it to zero:

$$\frac{\partial N}{\partial k} = n + n^{\frac{k+1}{k}} - k \ln n \frac{1}{k^2} n^{\frac{k+1}{k}} = n\left(1 + n^{\frac{1}{k}} - n^{\frac{1}{k}} \frac{\ln n}{k}\right) \stackrel{!}{=} 0.$$

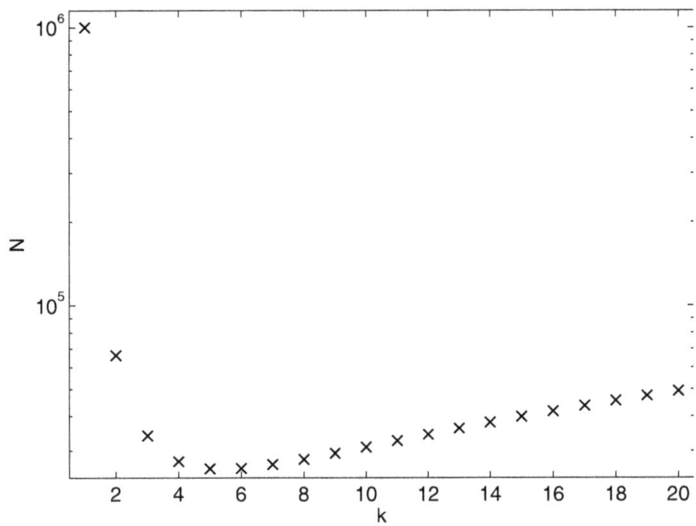

Fig. 4.2 The number of required units for a routing architecture between two layers depends strongly on the number $k-1$ of intermediate layers being used. The values shown here are for input and output stages of $n = 1000$ image points each.

This is satisfied for

$$\frac{\ln n}{k} = n^{-\frac{1}{k}} + 1. \tag{4.3}$$

With the ansatz

$$k_{opt} = c \ \ln n \tag{4.4}$$

(4.3) becomes independent of n:

$$\frac{1}{c} = e^{-\frac{1}{c}} + 1. \tag{4.5}$$

Solving this numerically we obtain

$$k_{\rm opt} \approx 0.7822 \ \ln n. \tag{4.6}$$

The fact that $k_{\rm opt}$ scales logarithmically with n is not surprising by itself. Such a scaling behavior lies at the heart of many techniques that have to permute or operate on a group of nodes simultaneously, like permutation networks or the fast Fourier transform (Cooley and Connor, 1965). Even in random graphs (Erdös and Rényi, 1959) the network diameter (corresponding somewhat to our number of layers k) scales logarithmically with the number of nodes. This general logarithmic scaling behavior is independent of the specific fanout (or degree) at each node. A different fanout only changes the basis of the logarithm, which is equivalent to changing the prefactor in the logarithmic relation. Here, however, minimizing the number of components of the network leads to a *specific* logarithmic scaling, or phrased differently: the prefactor c in k_{opt} is unique. This goes hand in hand with the existence of a unique optimal fanout

$$l_{\text{opt}} = n^{\frac{1}{k_{\text{opt}}}} = e^{\frac{1}{c}}. \tag{4.7}$$

We will discuss this finding further in Section 4.5.

More than One Feature Per Image Point

So far, we have neglected the routing of visual information when there are several feature cells at one image point. Instead of a dense pixel array, visual information in V1 is represented by a pattern of "hypercolumns" of lower density. However, each hypercolumn contains cells responsive to many different local properties of the input, such as wavelet-like features (Gabors) at different orientations and spatial frequencies, different colors or specificity for one eye or the other.

It may not be necessary to route these features independently of each other to higher areas, so one might assume that only one active link is needed to route many feature units in one image location. On the other hand, certain feature types do require individual treatment. For example, for full orientation invariance, units of one orientation specificity of the input would need connections to all orientation specificities of the output domain. The truth probably lies somewhere in between these two extremes, as suggested in (Zhu and von der Malsburg, 2004): image points are not routed individually, but in small assemblies through collective links called "maplets". For every group of nodes, there exist several such maplets, responsible for routing at different scales and orientations without requiring individual links for all features in all positions.

Since the focus of this analysis is not on a specific routing architecture, but on finding the optimal number of layers for a very general architecture, we will merge the above arguments into a single factor $\alpha \geq 1$ representing the number of feature nodes that are controlled by a single link. If necessary, the parameter α can also be used to account for unequal expense assumed for feature units versus link units.

Instead of n independent feature nodes, we now have

$$n_\alpha = \frac{n}{\alpha} \tag{4.8}$$

groups of nodes, each containing α nodes. With this, the number of units in the routing circuit (4.2) changes to

$$N(k,n) = \alpha(k+1)n_\alpha + kn_\alpha^{\frac{k+1}{k}}. \tag{4.2'}$$

Setting the derivative of (4.2') to zero leads to

$$\frac{\ln n_\alpha}{k} = \alpha n_\alpha^{-\frac{1}{k}} + 1 = 0. \tag{4.3'}$$

The new ansatz

$$k_{opt} = c \ln n_\alpha \tag{4.4'}$$

yields

$$\frac{1}{c} = \alpha e^{-\frac{1}{c}} + 1, \tag{4.5'}$$

which we can solve numerically for explicit values of α. In Figure 4.3 we see that c—and with it k_{opt}—only changes by a factor of 2 over a reasonably large range of α.

Having determined the number of layers k_{opt} that minimizes the required neural circuitry for given n and α, we can calculate the size N_{opt} of this minimal circuitry. Inserting (4.4') into (4.2') yields

$$N_{opt}(n) = n + \left(\alpha + e^{\frac{1}{c}}\right)(cn_\alpha \ln n_\alpha). \tag{4.9}$$

This means that for large n the number of units of the optimal routing architecture between two layers of n_α image points scales with

$$N_{opt}(n_\alpha) \propto n_\alpha \ln n_\alpha, \tag{4.10}$$

as expected from classical network theory. This result holds also for routing of only a single feature ($\alpha = 1$) per point.

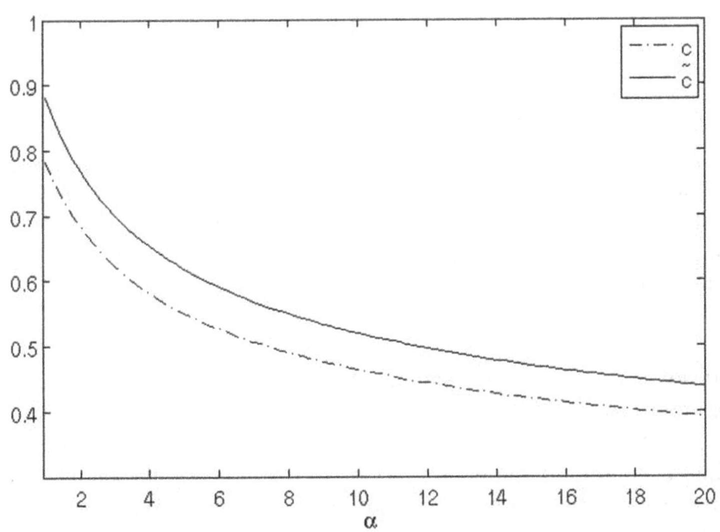

Fig. 4.3 The prefactors c and $\tilde{c}$ define k_{opt} through (4.4') and (4.13) in the cases of routing to an output of the same size or much smaller size, respectively.

4.3.2 Routing Circuit with Different Sizes of Input and Output Layer

Let us now discuss routing from the whole visual field to a comparatively small cortical output region. We assume that an attentional mechanism singles out, in the input domain, a region that is to be mapped to the output region.

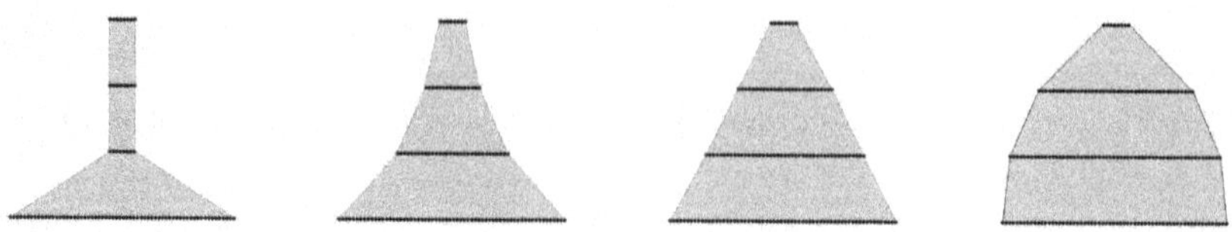

Fig. 4.4 Possible forms of tapered networks. All these ways of decreasing layer size towards the output are possible in principle. For our analysis we choose a linear decrease (center right) as the most parsimonious solution.

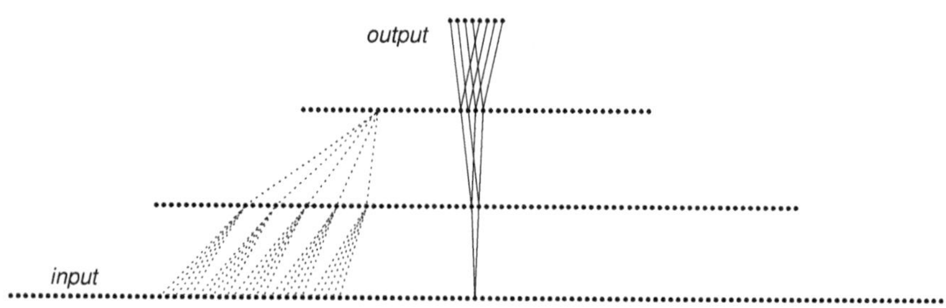

Fig. 4.5 Routing network for an input of $n = 125$ and an output of $\frac{n}{m} = 8$ nodes with $k = 3$ link layers and linear decrease of layer size. Consequently, the upward fan-out is $l = 2$. The full lines show the links connecting an input node with the full output stage. Downward connectivity is $l_{down} = n^{\frac{1}{3}} = 5 > l$ (shown exemplarily for a node on the second level of the architecture by dotted lines).

We do not claim here that invariant recognition is perfect over the whole visual field (there are studies showing that this is not the case, like Dill and Fahle; Cox et al., 1998; 2005), but object recognition *is* possible to some degree even at high retinal eccentricities, although of course impaired by the poor resolution at these parts of the retina. In any case, the basic problem remains the same as before: neural connections must exist between all parts of the visual input region and a target area.

Computationally, the situation is very similar to the one discussed in the previous section and leads to a generalization of the results derived there. We now want to connect an input stage of n units with an output stage that is smaller by the factor m and contains only $\frac{n}{m}$ units. As before, the routing is established via $k-1$ intermediate layers, and groups of α nodes can be routed collectively. With the same argument as in Section 4.3.1 we see that now each group has to make

$$l = \left(\frac{n\alpha}{m}\right)^{\frac{1}{k}} \tag{4.11}$$

connections to the next higher layer in order to connect every group of input nodes with every output group. Figure 4.5 shows parts of the architecture required for routing from a 125 node input to an 8 node output stage. Note that due to the different input and output sizes downward fan-out now has to be higher than the upward fan-out l.

Differently from before, the size of intermediate layers is not well-defined now. In principle an architecture is conceivable where already after the first routing stage the size of the feature layers is reduced to that of the output stage (Figure 4.4, left). Although this still allows all input units to be connected to all output points, it would require a comparatively high number of downward connections for the feature units in the first intermediate layer. Moreover, this decrease of layer sizes clearly contradicts anatomical reality. Other possibilities include geometric or linear decrease of intermediate layers (Figure 4.4, center) or the sub-linear tapering on the right side, which was used in (Olshausen et al., 1993). We will assume here that the number of nodes changes linearly from the input to the output layer. This is supported by measurements of the average sizes of primary visual areas in humans (Dougherty et al., 2003). Note, however, that the same paper reports variance of V1 sizes of more than 100% between different individuals. In general there seems to be little undisputed data on this question in the literature. Given this uncertainty in the anatomical data, the simplest possible assumption is probably best for this kind of general discussion.

With a linear decrease in size, the number of feature units in layer κ ($\kappa = 0$ for the input and $\kappa = k$ for the output layer) is

$$f_\kappa = n - \frac{\kappa}{k}\left(n - \frac{n}{m}\right).$$

The number F of the feature encoding units of all layers is then

$$F = \sum_{\kappa=0}^{k} f_\kappa = (k+1)n - \left(n - \frac{n}{m}\right)\frac{k(k+1)}{2k} = \frac{n}{2}\frac{m+1}{m}(k+1).$$

Adding the links emanating upwards from all but the top-most layer, we get the total number of units as

$$\begin{aligned} N(n,k) &= F + \frac{1}{\alpha}\left(F - \frac{n}{m}\right) l \\ &= \frac{n\alpha}{2}\frac{m+1}{m}\left[\alpha(k+1) + \left(k+1-\frac{2}{m+1}\right)\left(\frac{n\alpha}{m}\right)^{\frac{1}{k}}\right]. \end{aligned} \tag{4.12}$$

Setting the derivate with respect to k to zero leads to

$$-\alpha\left(\frac{n_\alpha}{m}\right)^{-\frac{1}{k}} = 1 + \frac{\ln\frac{n_\alpha}{m}}{k^2}\left(\frac{2}{m+1} - k - 1\right).$$

With the ansatz

$$k_{opt} = \tilde{c}\ln\frac{n_\alpha}{m} \tag{4.13}$$

this turns into

$$-\alpha e^{-\frac{1}{\tilde{c}}} = 1 + \frac{1}{\tilde{c}^2\ln\frac{n_\alpha}{m}}\left(\frac{2}{m+1} - 1\right) - \frac{1}{\tilde{c}}.$$

For large input/output ratio m, the term $\frac{2}{m+1}$ becomes negligible, so that $\tilde{c}$ depends only on the number of independently routed output nodes $\frac{n_\alpha}{m}$ and not on m itself:

$$-\alpha e^{-\frac{1}{\tilde{c}}} \approx 1 - \frac{1}{\tilde{c}^2\ln\frac{n_\alpha}{m}} - \frac{1}{\tilde{c}}. \tag{4.14}$$

Numerical analysis of (4.14) shows, however, that $\tilde{c}$ changes by less than 10% when $\frac{n_\alpha}{m}$ is varied over 3 orders of magnitude. So we can say that, like c in Section 4.3.1, $\tilde{c}$ only depends on the parameter α. Figure 4.3 shows that $\tilde{c}$ takes on similar but slightly higher values than c.

Calculating the size of the derived routing circuit by plugging (4.13) into (4.12) yields

$$N_{opt} = \left(\alpha + e^{\frac{1}{\tilde{c}}}\right)\frac{n_\alpha}{2}\left(\tilde{c}\ln\frac{n_\alpha}{m} + 1\right) \tag{4.15}$$

for large m. Although the relation is a bit different from the one derived for equal input and output domains (4.9), the scaling with $n_\alpha \ln n_\alpha$ (since $n_\alpha \gg m$) remains the same.

4.4 Interpretation of Results

Let us now discuss functional and physiological implications of the architecture derived in Section 4.3.

4.4.1 Difference to Sorting Networks

sorting network It is important to note that the routing architectures derived here are different from sorting networks known in computer science. A sorting network takes a number of scalar input signals and outputs them at the final stage ordered by their values. At each intermediate node, (usually) two signals converge and are compared by a logic unit to be passed on in an ordered fashion. Switchyards, on the other hand, shift and reorganize signals by a scheme imposed from outside (via the control units), without internally comparing signals. The intermediate nodes ($\widehat{=}$ feature units) do not have any active role but are pure relay stations that combine any signals they receive and pass them on (this may be a reason to use max-like operations at the feature units in future models). Therefore, Switchyards run the risk of having *conflicts* in their routing, which is impossible by definition for sorting networks. A conflict occurs whenever the signal at two or more feature nodes at one stage is sent to the same feature node of the next stage. Here, the signals are inevitably mixed and cannot be separated any more. Note, however, that a conflict is problematic only if it mixes two totally unrelated signals. In some cases, e.g. when downscaling visual information, "conflicts" are unavoidable and actually desirable, since in this case high resolution information of neighboring points needs to be averaged ($\widehat{=}$ mixed) into a single point.

The question in which situations conflicts arise in routing networks has not been investigated in depth so far. For Switchyards, no conflicts occur if pure shift or rotation operations are carried out. Scalings produce conflicts exactly if the scaling factor is a multiple of the fanout l of the Switchyard. Whether a conflict can occur between the paths emanating from two input nodes also depends on their proximity. Figure 4.6 shows the number of conflicts that can occur between two nodes, scaled by the absolute number of possible paths that can be taken. While for two directly neighboring input nodes about 24% of possible choices of output nodes will lead to conflicts somewhere along the path, this number decreases to 0 for very distant nodes if we discount the degenerate case of identical output nodes. More precisely speaking, paths emanating from input nodes with a distance

$$d \geq \frac{n}{l} \tag{4.16}$$

can never collide no matter what output nodes are chosen (assuming the output nodes are not identical). Most conflicts in Switchyards will anyway arise from the paths of two neighboring input nodes merging already at the first routing stage. This is not very problematic since owing to the nature of visual information, it does not mix totally independent signals, but signals that are very related anyway. Thus the merging merely acts as a kind of low pass filter. This kind of low pass behavior is even necessary to perform correspondence finding in switchyards, as we will see in Chapter 6.

Although Switchyards are well suited for routing of visual information, sorting networks outperform them in raw power of re-routing since they can realize any kind of input-output mapping. One reason for this lies in the aforementioned fact that their intermediate nodes are active logical units instead of passive relay nodes. Moreover, sorting networks require larger circuits for given input and output sizes. Our Switchyards are of size $\mathcal{O}(n \log n)$ with a relatively small prefactor. Functional sorting networks, on the other hand, scale at least with $\mathcal{O}(n(\log n)^2)$, more common and easier to implement ones even with $\mathcal{O}(n^x)$ with $x = 1.5$ or higher (for an extensive treatment, see Knuth, 1997). Ajtai et al. (1983) describe a sorting network of order $\mathcal{O}(n \log n)$. However, this network is not used in practice since the prefactor in the size is huge and the explicit formulation of quite complicated expander graphs would be required.

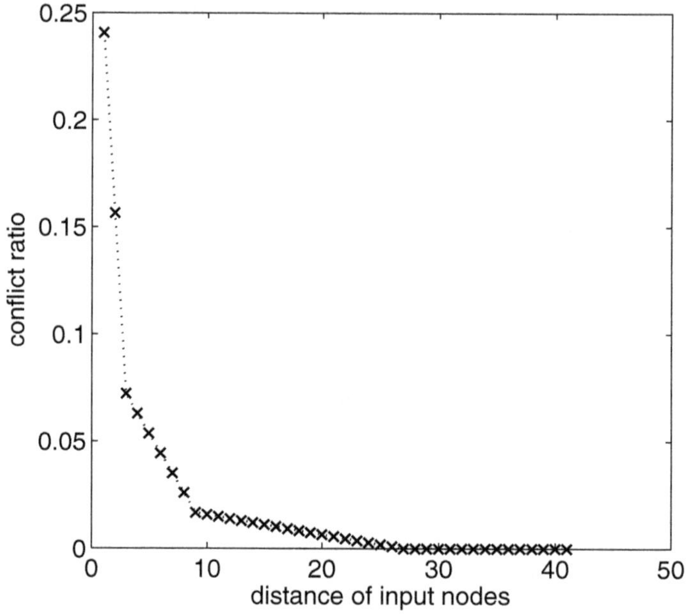

Fig. 4.6 Number of possible conflicts as a function of distance of input nodes. Shown here for a network with $n = 81$ input and output nodes, fanout $l = 3$, and $k = 4$ routing stages. The abscissa marks the separation of two input nodes from which all possible pathways to arbitrary but different output nodes are evaluated. The absolute position of the input nodes is irrelevant since Switchyards are symmetric to shifts. On the ordinate, we plot the number of pathway constellations that lead to conflicts, relative to the overall number of possible pathway combinations. For a distance larger or equal than $\frac{n}{l}$ ($= 27$ for this case), no conflicts can occur any more.

We conclude that sorting networks and Switchyards are not directly comparable. While the former can re-arrange signals arbitrarily, they have to use active logic elements at all relay nodes and are relatively large. Switchyards, on the other hand, have limited routing capabilities which are somewhat tailored for visual information routing, and they only scale with $\mathcal{O}(n \log n)$.

4.4.2 *Physiological Interpretation*

In Section 4.3 we found that the optimal number of link layers in a routing circuit is given by

$$k_{opt} = c \ \ln n_{\alpha}$$

and

$$k_{opt} = \tilde{c} \ln \frac{n_{\alpha}}{m}$$

for routing to an output stage of identical size and of much smaller size, respectively. So in both cases, k_{opt} is proportional to the natural logarithm of the number of independently routed *output* nodes. The well defined prefactors c and $\tilde{c}$ are very similar (cf. Figure 4.3), depend only on α, and do not vary too much over large ranges of α.

How do those results match the facts in the human brain? A good starting point is the optic nerve, which is known to contain $\sim 10^6$ fibers for humans and other primates (Potts et al., 1972). Since the optic nerve is the bandwidth bottleneck of the visual system, it is safe to assume that it contains no redundant information. The number of neurons in V1, however, is by far higher than the number of optic nerve fibers, mainly for two reasons. First, the cortex most probably employs a population coding strategy in order to reduce noise and increase transmission speed. This means that several neurons together (perhaps the ≈ 100 of a cortical minicolumn, cf. Section 3.3) represent one of our abstract feature units. Second, visual information is represented in an overcomplete code in V1 (Olshausen and Field, 1997), increasing the number of feature units over the number of optic nerve fibers. Nevertheless, the information represented in V1 cannot be higher than that transported by the optic nerve, so that overcomplete groups of units can be routed collectively. We will therefore assume that the number of feature encoding units is of the same order as the number of fibers in the optic nerve, keeping in mind that an overcomplete basis in V1 may be accounted for by a correspondingly higher value of α.

From the primary visual area V1, visual information is routed retinotopically along the ventral pathway to a target region in inferotemporal cortex (IT). Psychophysical evidence (van Essen et al., 1991) suggests that about 1000 feature nodes are sufficient to represent the contents of the two dimensional "window of attention", and therefore the size of this target region, at any given time. One may assume that there exist multiple such target regions in parallel in IT, which are used for different object recognition tasks (we sketch this idea in Chapter 7).

How would our routing architecture look for these numbers? For this, we still miss an estimate of the parameter α. Research in our lab has shown that representing an image with 40 Gabor wavelets in each image point preserves all necessary image information of gray scale images (Wundrich et al., 2004) and allows good object identification (Lades et al., 1993). To additionally include color and temporal information (direction of motion), this number would have to be roughly twice as high. This is in line with findings concerning the number of orientation pinwheels in the primate brain. (Obermayer and Blasdel, 1997) report around 10^4 pinwheels for V1 of the Macaque. Assuming a similar number for the human brain, we face the situation of an input region of the ventral stream containing 10^6 feature units clustered in some 10^4 pinwheels. If we assume that every pinwheel—as a first order approximation of the functional "hypercolumn"—contains the full set of visual features for a certain input location on one retina, it follows that the number of these distinct features is of the order 100. Inputs from the two eyes are treated independently here, so that successful stereoscopic fusion can be achieved for arbitrary depths by activating the right routing links.

While we have two agreeing estimates of the number of feature units per resolution point, coming from computer vision and physiology, the number α of features that can be routed together is difficult to estimate. It depends on the kinds of invariance operations that are actually realized in the routing circuit, as discussed in the second part of Section 4.3.1. We assume α to lie in the approximate range of 2 to 5. For these values, the optimal number of layers for routing $(k+1)$ from a 10^6 node input to a 1000 node output ranges from 4.3 to 5.8. Figure 4.7 shows these values, as well as the number of units required for the full circuit when using the optimal number of layers.

The ventral pathway comprises the areas V1, V2, V4, and IT. IT in turn consists of posterior, central, and anterior parts. In our setting it may make sense to take into account this additional subdivision, since the receptive field sizes of these three parts are very different (Tanaka et al., 1993; see also Figure 4 in Oram and Perret, 1994), suggesting that they form different stages of the routing hierarchy. Visual information is relayed from the lateral geniculate nucleus in a rather clear sequential order V1→V2→V4→PIT→CIT→AIT, finally being combined with other signal streams in the superior temporal polysensory area (STP). Note that there exist at least equally

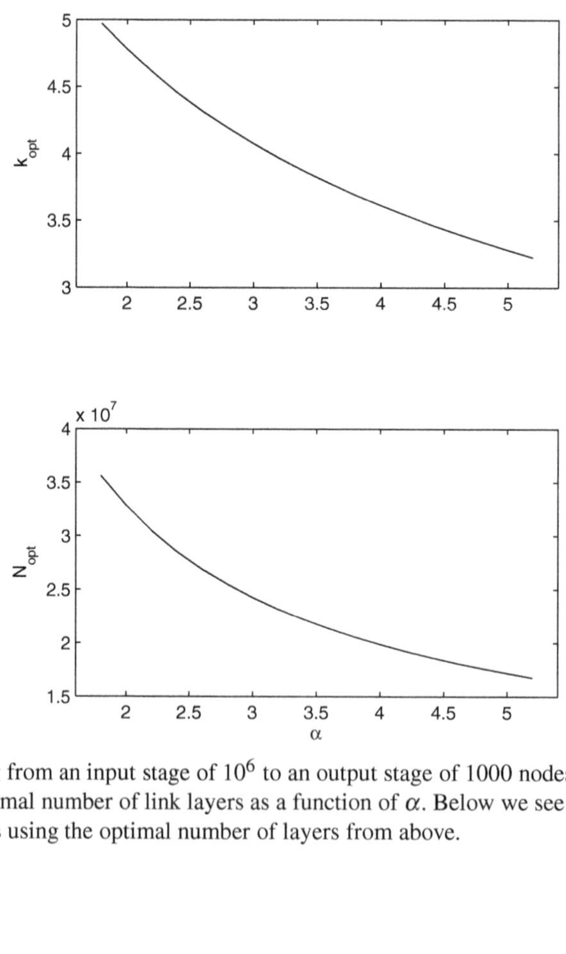

Fig. 4.7 Routing from an input stage of 10^6 to an output stage of 1000 nodes. The upper part displays the optimal number of link layers as a function of α. Below we see the total number of required units using the optimal number of layers from above.

strong feed*back* connections between the layers, indicating the importance of recurrent processes in vision. The number of 4–6 distinct cortical stages (depending on whether we regard IT as one or three stages) lies clearly in the range derived for our optimal circuit above. It is therefore possible that the ventral pathway indeed performs computationally optimal information routing. At this point, however, this is only a hypothesis, due to the great uncertainties in the available data. More explicit interpretations would be possible if α could be narrowed down further (also by quantifying the "overcompleteness" of V1) and if the stages involved in the routing were known for certain. Also, more psychophysical work on the information content in the window of attention would be desirable.

4.5 Discussion

We have seen in Section 4.3 that under some very general assumptions there exists a clear optimality condition on the number of layers required to build a routing architecture with minimal neural resources. This number depends on the size of the target region as well as the number of independently routed feature types. Within the given uncertainties, the derived numbers agree well with physiological data.

Constraining the design of a routing architecture by an optimality condition, as we did, has the advantage of imposing an additional requirement to an otherwise underconstrained problem. While the Shifter Circuit of Olshausen et al. (1993) addresses several anatomical and physiological facts, there is no experimental or theoretical justification for some of the parameter values chosen, among them the exact doubling of link spacing from layer to layer. In the absence of experimental results dictating these values, we think it best to follow some global optimality condition like proposed above.

In this chapter we have taken very general assumptions, and as few as possible. Of course, there are several plausible ways in which these assumptions can be made more specific, possibly changing the derived quantitative results:

- We avoided on purpose a detailed discussion of the kind of feature-to-feature connectivity that may be in place to achieve scale and orientation invariance. This is being addressed in ongoing work at our institute and will help to narrow down the parameter α.
- Routing architectures with many numbers of layers are a disadvantage to the organism in terms of longer reaction times and more complicated routing dynamics. This additional cost has not been considered here, its influence would bias the biological routing architecture in favor of fewer stages than derived here.

The analysis carried out in Section 4.3 also leads to experimental predictions. One such prediction arises from the fact that the notion of a static receptive field becomes meaningless if one embraces an active routing process. During attention focusing and recognition, this process would choose a certain routing path and deactivate all alternative pathways. For a unit at the output stage in the hierarchy (IT), this would change the functional receptive field from a very broad region to a narrow and specific location. A unit at a medium stage of the hierarchy might even be bypassed by the currently established routing pathway. There is ample evidence for the behavioral plasticity of receptive fields (Moran and Desimone, 1985; Connor et al., 1993), and recent findings (Murray et al., 2006) show that even the size of representation in V1 can change with an object's perceived size (suggesting a scale invariant routing process that already starts in the mapping from LGN to V1). However, these findings are often interpreted as the result of a diffuse "attention modulation" mechanism, without taking the possibility of an explicit routing process seriously. In the light of the rather specific geometric changes of receptive fields implied by the presence of such a process, it should be possible to design attention experiments that can clearly prove or refute the routing hypothesis.

While the above predictions are general implications of any multistage routing process and have been discussed similarly before (Olshausen et al., 1993), the quantitative results obtained here make some more specific predictions. An interesting feature of the minimal architecture, already mentioned in Section 4.3.1, is that the number of links emanating from one node (see (4.1) or (4.11)) is independent of network size:

$$l_{\text{opt}} = n_\alpha^{\frac{1}{k_{\text{opt}}}} = \exp\left(\frac{\ln n_\alpha}{\tilde{c}\ln n_\alpha}\right) = \exp\left(\frac{1}{\tilde{c}}\right). \tag{4.17}$$

l_{opt} is surprisingly low (between 3 and 9 for the range of α shown in Figure 4.3). This number should not be confused with the full number of connections that a cortical neuron makes, which is known to be several thousand. First, here we only count the connections necessary for information routing, not those involved in other kinds of processing or communication (e.g., feature extraction). Second, as mentioned above, the functional units discussed here are abstract "image points", which in the cortex are probably made up of ≈ 100 spiking neurons (like a cortical minicolumn). Single neurons in such a group would have to devote the majority of their connections to homeostatic within-group connections (cf. Lücke and von der Malsburg, 2004, and see the discussion in Sect. 3.3), which do not appear on our level of abstraction. Nevertheless, the small fanout necessary for optimal routing is an interesting feature and shows that by including the number of control units into our optimization we have implicitly also minimized the required connectivity of the routing architecture.

The optimal number of layers (equations 4.4' and 4.13), on the other hand, scales logarithmically with network size:

$$k_{opt} = \tilde{c}\ln n_\alpha.$$

This means that if more visual information has to be routed, the number of routing stages increases, while the local properties (number of connections that each node has to make) remain the same. Consequently, for species processing different amounts of visual information, the ventral streams should contain different numbers of routing stages. While the optic nerve of primates contains on the order of 10^6 fiber (Potts et al., 1972), the number is 10^5 for the rat (Fukuda et al., 1982), $2 \cdot 10^5$ for the cat (Hughes and Wässle, 1976), and $2.4 \cdot 10^6$ for the adult chicken (Rager and Rager, 1978). If we assume, as we did before, that the number of optic nerve fibers is a measure for the number of input units of the ventral stream, *and* if the number of output (IT) units changes by the same factor, then a rat would optimally have 2.3 layers less, a cat 1.6 layers less, and a chicken 0.9 routing layers more than a primate. The differences might be smaller, however, if the size of the output stage does not change as strongly as the number of optic nerve fibers, since k_{opt} depends on the number of output units. Although anatomical comparisons across species will be difficult, it may be interesting to investigate different brains with regard to this question.

Although the Switchyards derived in this chapter are useful for visual information routing and may be optimal in some sense, looking at those patterns provokes the question "how can such complex structures possibly arise in the brain?". Let us turn to this question in the following chapter.

References

Ajtai, M., Komlós, J., Szemerédi, E.: An 0(n log n) sorting network. In: Proceedings of the fifteenth annual ACM symposium on Theory of computing, pp. 1–9 (1983)

Cherniak, C.: The bounded brain: Toward quantitative neuroanatomy. Journal of Cognitive Neuroscience 2(1), 58–68 (1990)

Connor, C.E., Gallant, J.L., Preddie, D.C., van Essen, D.C.: Responses in area v4 depend on the spatial relationship between stimulus and attention. Journal of Neurophysiology 75, 1306–1308 (1993)

Cooley, J.W., Connor, J.W.C.E.: An algorithm for the machine calculation of complex fourier series. Mathematics of Computation 19, 297–301 (1965)

Cox, D., Meier, P., Oertelt, N., DiCarlo, J.J.: 'breaking' position-invariant object recognition. Nature Neuroscience 8(9), 1145–1147 (2005)

Desimone, R., Wessinger, M., Thomas, L., Schneider, W.: Attentional control of visual perception: cortical and subcortical mechanisms. Cold Spring Harb. Symp. Quant. Biol. 55, 963–971 (1990)

Dill, M., Fahle, M.: Limited translation invariance of human visual pattern recognition. Perception and Psychophysics 60(1), 65–81 (1998)

Dougherty, R.F., Koch, V.M., Brewer, A.A., Fischer, B., Modersitzki, J., Wandell, B.A.: Visual field representations and locations of visual areas v1/2/3 in human visual cortex. Journal of Vision 3, 586–598 (2003)

Duhamel, J., Bremmer, F., BenHamed, S., Graf, W.: Spatial invariance of visual receptive fields in parietal cortex neurons. Nature 389, 845–848 (1997)

Durbin, R., Rumelhart, D.E.: Product units: A computationally powerful and biologically plausible extension to backpropagation networks. Neural Computation 1(1), 133–142 (1989)

Elston, G.N., Rosa, M.G.: Morphological variation of layer iii pyramidal neurones in the occipitotemporal pathway of the macaque monkey visual cortex. Cereb Cortex 8(3), 278–294 (1998)

Elston, G.N., Rosa, M.G.: Pyramidal cells, patches, and cortical columns: a comparative study of infragranular neurons in teo, te, and the superior temporal polysensory area of the macaque monkey. J Neurosci. 20(24), RC117 (2000)

Erdös, P., Rényi, A.: On random graphs. Publicationes Mathematicae 6, 290–297 (1959)

Freund, T., Bolam, J., Björklund, A., Stenevi, U., Dunnett, S., Powell, J., Smith, A.: Efferent synaptic connections of grafted dopaminergic neurons reinnervating the host neostriatum: a tyrosine hydroxylase immunocytochemical study. J. Neurosci. 5, 603–616 (1985)

Fukuda, Y., Sugimoto, T., Shirokawa, T.: Strain differences in quantitative analysis of rat optic nerve. Experimental neurology 75, 525–532 (1982)

Gabbiani, F., Krapp, H.G., Koch, C., Laurent, G.: Multiplicative computation in a visual neuron sensitive to looming. Nature 420, 320–324 (2002)

Hughes, A., Wässle, H.: The cat optic nerve: Fibre total count and diameter spectrum. The Journal of Comparative Neurology 169, 171–184 (1976)

Hung, C.P., Kreiman, G., Poggio, T., DiCarlo, J.J.: Fast readout of object identity from macaque inferior temporal cortex. Science 310(5749), 863–866 (2005), http://dx.doi.org/10.1126/science.1117593

Knuth, D.: The Art of Computer Programming, vol. 3, ch. 5.3.4: Networks for Sorting. Addison-Wesley, Reading (1997)

Koch, C.: Biophysics of computation: information processing in single neurons. Oxford University Press, Oxford (1999)

Konen, C.S., Kastner, S.: Two hierarchically organized neural systems for object information in human visual cortex. Nature Neuroscience 11(2), 224–231 (2008), http://dx.doi.org/10.1038/nn2036

Kubota, Y., Hatada, S., Kondo, S., Karube, F., Kawaguchi, Y.: Neocortical inhibitory terminals innervate dendritic spines targeted by thalamocortical afferents. J Neurosci. 27(5), 1139–1150 (2007) (comparative Study)

Lades, M., Vorbrüggen, J., Buhmann, J., Lange, J., von der Malsburg, C., Würtz, R., Konen, W.: Distortion invariant object recognition in the dynamic link architecture. IEEE Transactions on computers 42, 300–311 (1993)

Larkum, M., Zhu, J., Sakmann, B.: A new cellular mechanism for coupling inputs arriving at different cortical layers. Nature 398, 338–341 (1999)

Lücke, J., von der Malsburg, C.: Rapid processing and unsupervised learning in a model of the cortical macrocolumn. Neural Computation 16, 501–533 (2004)

Lund, J.S., Yoshioka, T., Levitt, J.B.: Comparison of intrinsic connectivity in different areas of macaque monkey cerebral cortex. Cereb Cortex 3(2), 148–162 (1993)

Moran, J., Desimone, R.: Selective attention gates visual processing in the extrastriate cortex. Science 229, 782–784 (1985)

Murray, S.O., Boyaci, H., Kersten, D.: The representation of perceived angular size in human primary visual cortex. Nature Neuroscience 9, 429–434 (2006)

Nicola, S., Surmeier, J., Malenka, R.: Dopaminergic modulation of neuronal excitability in the striatum and nucleus accumbens. Annu. Rev. Neurosci. 23, 185–215 (2000)

Obermayer, K., Blasdel, G.G.: Singularities in primate orientation maps. Neural Computation 9, 555–575 (1997)

Olshausen, B.A., Anderson, C.H., van Essen, D.C.: A neurobiological model of visual attention and invariant pattern recognition based on dynamic routing of information. Journal of Neuroscience 13(11), 4700–4719 (1993)

Olshausen, B.A., Field, D.J.: Sparse coding with an overcomplete basis set: a strategy employed by v1? Vision Research 37, 3311–3325 (1997)

Oram, M.W., Perret, D.I.: Modeling visual recognition from neurobiological constraints. Neural Networks 7, 945–972 (1994)

Perkel, D.J., Bullier, J., Kennedy, H.: Topography of the afferent connectivity of area 17 in the macaque monkey: A double-labelling study. The Journal of Comparative Neurology 253, 374–402 (1986)

Pitts, W., McCulloch, W.S.: How we know universals: the perception of auditory and visual forms. Bulletin of Mathematical Biophysics 9, 127–147 (1947)

Postma, E., van den Herik, H., Hudson, P.: SCAN: A Scalable Model of Attentional Selection. Neural Netw. 10(6), 993–1015 (1997)

Potts, A.M., Hodges, D., Shelman, C.B., Fritz, K.J., Levy, N.S., Mangnall, Y.: Morphology of the primate optic nerve. i. method and total fiber count. Investigative Ophthalmology & Visual Science 11, 980–988 (1972)

Pouget, A., Sejnowski, T.J.: Spatial transformations in the parietal cortex using basis functions. Journal of Cognitive Neuroscience 9(2), 222–237 (1997)

Rager, G., Rager, U.: Systems-matching by degeneration i. a quantitative electron microscopic study of the generation and degeneration of retinal ganglion cells in the chicken. Experimental Brain Research 33, 65–78 (1978)

Salinas, E., Sejnowski, T.J.: Gain modulation in the central nervous system: where behavior, neurophysiology, and computation meet. Neuroscientist 7(5), 430–440 (2001)

Schaefer, A.T., Larkum, M.E., Sakmann, B., Roth, A.: Coincidence detection in pyramidal neurons is tuned by their dendritic branching pattern. J Neurophysiol. 89(6), 3143–3154 (2003), http://dx.doi.org/10.1152/jn.00046.2003

Schwartz, E.L.: Spatial mapping in primate sensory projection: analytic structure and relevance to perception. Biological Cybernetics 25, 181–194 (1977)

Tal, D., Schwartz, E.L.: Computing with the leaky integrate-and-fire neuron: Logarithmic computation and multiplication. Neural Computation 9, 305–318 (1997)

Tanaka, K., Fujita, I., Kobatake, E., Cheng, K., Ito, M.: Serial processing of visual object-features in the posterior and anterior parts of the inferotemporal cortex. In: Ono, T., Squire, L.R., Raichle, M.E., Perrett, D., Fukuda, M. (eds.) Brain mechanisms of perception and memory: From neuron to behaviour, pp. 34–46. Oxford University Press, Oxford (1993)

Tanigawa, H., Wang, Q., Fujita, I.: Organization of horizontal axons in the inferior temporal cortex and primary visual cortex of the macaque monkey. Cerebral Cortex 15, 1887–1899 (2005)

van Essen, D.C., Olshausen, B., Anderson, C.H., Gallant, J.: Pattern recognition, attention, and information bottlenecks in the primate visual system. In: Mathur, B., Koch, C. (eds.) Proceedings of the SPIE Conference on Visual Information Processing: from neurons to chips, vol. 1473, pp. 17–28 (1991)

Volgushev, M., Vidyasagar, T., Pei, X.: A linear model fails to predict orientation selectivity of cells in the cat visual cortex. J. Physiol (Lond.) 496(Pt 3), 597–606 (1996)

Wolfrum, P.: Information Routing, Correspondence Finding, and Object Recognition in the Brain, PhD thesis, Goethe University Frankfurt (2008)

Wolfrum, P., von der Malsburg, C.: What is the optimal architecture for visual information routing? Neural Computation 19(12), 3293–3309 (2007), http://dx.doi.org/10.1162/neco.2007.19.12.3293

Wundrich, I.J., von der Malsburg, C., Würtz, R.P.: Image representation by complex cell responses. Neural Computation 16(12), 2563–2575 (2004), http://dx.doi.org/10.1162/0899766042321760

Zhu, J., von der Malsburg, C.: Maplets for correspondence-based object recognition. Neural Networks 17(8-9), 1311–1326 (2004)

Zipser, D., Andersen, R.A.: A back-propagation programmed network that simulates response properties of a subset of posterior parietal neurons. Nature 331(6158), 679–684 (1988), http://dx.doi.org/10.1038/331679a0

Chapter 5
Ontogenesis of Switchyards

In the previous chapter, we derived an optimized architecture for multistage routing of visual information. In order to be optimal in terms of number of neural elements, this network needs to have a very specific connectivity (cf. Figure 4.1). If we want to argue that such architectures are actually employed by the brain for information routing (which we do!), we must be able to explain how they develop in an animal.

One obvious assumption might be that this connectivity is directly encoded in the genes. For this to be possible, however, the raw information of the genome is too small. The human brain, for example, contains on the order of 10^{11} neurons. If we assume that each of them is only connected to 100 other neurons, describing this connectivity alone would require more than $4 \cdot 10^{13}$ bytes (40 terabytes). This is much more than the less than 1 gigabyte of information contained in the human genome (approx. 3200 megabases (Morton, 1991) with 2 bit each). Consequently, the structure of the brain and most likely also single routing networks cannot be directly encoded in the brain, but must emerge during development of the organism. Appreciation of this necessity leads to the search for principles of self-organization that can explain the emergence of complex structures from very few and compact mechanisms.

Studying self-organizational processes in the brain also has another, more indirect advantage. Sometimes researchers propose sophisticated models that can do "universal computation" or are "as powerful as a Turing machine". But although interesting in their own right, these models will not provide insight into how the brain works if there is no explanation of how they could develop in a living organism. In that sense the study of self-organization helps the search for candidate models of brain function by providing constraints to the otherwise large space of possible models that could in principle explain a certain function.

Models of development of brain function can either be based on autonomous mechanisms that do not require external information, or they can incorporate learning processes where sensory input shapes the neural circuits. While many models of the development of visual function rely on learning processes happening postnatally, we argue here that basic visual functions like content-independent routing of information should already be operational at birth. The newborns of many hoofed

P. Wolfrum: Info. Routing, Corres. Finding, and Object Recognition in the Brain, SCI 316, pp. 91–108.
springerlink.com © Springer-Verlag Berlin Heidelberg 2010

animals, for example, can stand up and walk around directly after birth, indicating that at least rudimentary orientation in visual space is already up and running.

We therefore propose here an ontogenetic mechanism for the development of routing structures like the Switchyards described in the previous chapter. We start out with a short overview of growth mechanisms in the brain (Section 5.1), before we proceed to describe the proposed model in detail (Section 5.2). Section 5.3 shows results and analyzes the behavior of the model. We discuss other potential mechanisms for the growth of Switchyards (Section 5.4) before concluding with Section 5.5. Parts of the results discussed in this chapter were previously published in (Wolfrum and von der Malsburg, 2007).

5.1 Ontogenetic Plasticity Mechanisms in the Brain

Over the last 50 years, a lot of interest has focussed on the mechanisms that shape connectivity structures in the brain. Basically, we can distinguish two categories of mechanisms.

- Chemical interaction between neurons and also among fibers influences the direction and speed at which axonal fibers grow.
- Once contact has been established between two neurons, their connections can grow or shrink depending on the firing activity of both neurons.

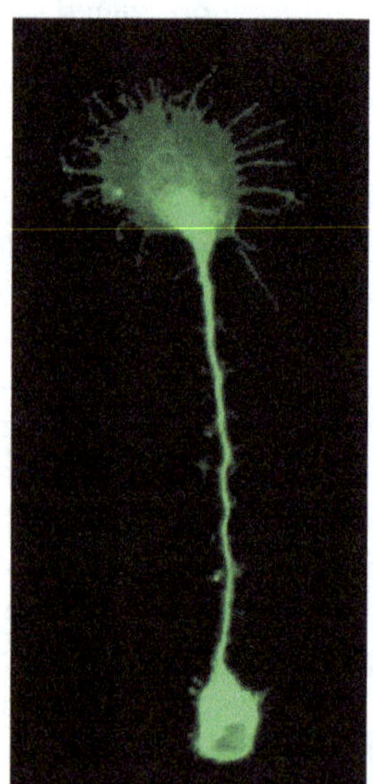

Fig. 5.1 Image of a rat superior cervical ganglion neuron with an axon growing upwards. The axon ends in a growth cone which regulates future direction and strength of growth. Diameter of the growth cone is approximately $10\mu m$. Courtesy of Zac Pujic and Geoff Goodhill.

An early idea for chemical interaction, the *chemoaffinity hypothesis* by Sperry (1963), assumes that axonal projections have a way of uniquely identifying their dedicated target location in the brain area they are growing towards. More recent findings, however, suggest that it is rather gradients of chemical markers that guide axon growth. When growing, an axon produces a *growth cone* with many little protrusions called *filopodia*. These filopodia can sense the gradient of chemical concentrations in their vicinity and will let the whole growth cone grow towards or withdraw from the source depending on whether the chemical has an attractive of repulsive effect on them (see Figure 5.1). For example, the retino-tectal projection in chicken is thought to be shaped by two opposing gradients of the substances "EphrinB" and "Wnt3" (Schmitt et al., 2006). Another chemical mechanism is provided by attraction or repulsion of neighboring fibers (e.g., Holt and Harris, 1993). Repulsive forces between fibers may be helpful in spreading connections homogeneously over the whole target tissue, while attractive forces could support the growth of parallel connections. See Table 5.1 for a non-exhaustive collection of substances involved in axon guidance.

Table 5.1 A sample of chemical substances involved in axon guidance. Geoff Goodhill, personal communication.

Ligand(s)	Receptor(s)	Functional Roles
Netrins / Unc6	DCC/Unc40	Attracts commissural axons to floorplate
	Unc5	Repels trochlear/cranial motor neurons
	Neogenin	Attracts retinal ganglion axons to optic disk Thalamocortical / corticothalamic Cell Migration
Semaphorins	Neuropilins	Repulsion of sympathetic and sensory axons
	Plexins	Cortical axon/dendrite guidance
Slits	Robos	Repulsion from midline Axonal branching
Ephrins	Eph	Map formation: retinotectal, hippocampal-septal Nervous system segmentation/patterning

Activity dependent mechanisms, on the other hand, play a role in refining connectivity once an axon has made contact to a target neuron. Most likely, they follow similar principles as the plasticity mechanisms that can be observed in the mature brain. These include the Hebb rule, anti-Hebbian, homeostatic mechanisms, or spike timing dependent plasticity. Experiments have shown that darkness or the blocking of neural activity by TTX (Olson and Meyer, 1991) or disturbing correlated firing with stroboscopic light (Schmidt and Buzzard, 1993) result in altered topographic projections, underlining the relevance of activity-based mechanisms. Consequently, for many years there was the notion that activity dependent mechanisms were required during ontogeny to produce the correct fine-structure of connections. However, recent findings suggest that under some circumstances chemical mechanisms alone may be sensitive enough. For example, Rosoff et al. (2004) have shown that growth cones can detect concentration differences as small as a single molecule across their spatial extent.

Below, we present an ontogenetic mechanism that is purely based on interactions of chemical markers. Recent findings (Hattori et al., 2009) support the assumptions made by this model (see the discussion in Section 5.4). However, we are more interested in the functional properties of the mechanism than claiming a specific biological implementation, and in Section 5.4 we show how the same growth behavior can be achieved using only activity-induced plasticity.

5.2 A Model for the Growth of Routing Networks

As we discussed in Chapter 4, Switchyards connect a layer of n input nodes via k routing stages to an output layer of $\frac{n}{m}$ nodes (see Figure 5.2). How can ontogeny produce such routing circuits in the brain? Especially the large gaps necessary between links on higher stages are difficult to explain with traditional learning rules. We will investigate below whether chemical markers can help forming such structures. For simplicity and ease of visualization we restrict ourselves here to the case of one-dimensional feature layers. However, the mechanism derived below can also create three-dimensional networks connecting two-dimensional sheets of neurons, as we see in Section 5.3.2.

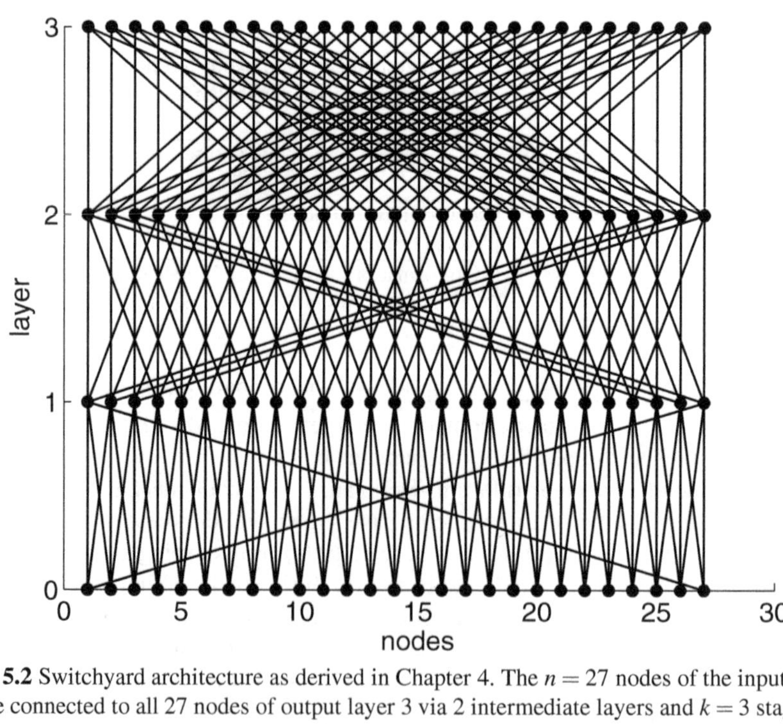

Fig. 5.2 Switchyard architecture as derived in Chapter 4. The $n = 27$ nodes of the input layer 0 are connected to all 27 nodes of output layer 3 via 2 intermediate layers and $k = 3$ stages of links.

Let $C^{\kappa}_{i,j}$ denote the strength of the link between node i in layer κ to node j of layer $\kappa+1$. $C^{\kappa}_{i,j}$ can vary between 0 and 1, with 0 representing an absent link and 1 a fully grown one. We will refer to all links of one stage by the $n \times n$ matrix C^{κ}. When we make a statement that refers to the links of all k stages we will leave out the superscript κ (this also applies for other variables introduced below).

We describe the growth of the links not directly in terms of C but of an unbounded variable U, which codes for the real links via the sigmoid function

$$C = \frac{1}{1+e^{-sU}}\ , \tag{5.1}$$

where s defines the steepness of the sigmoid. We let U start out at a homogeneous negative value with some noise added (see Section 5.3 for a discussion of robustness to noise), so that all links C are initially close to 0. The growth of U then follows the differential equation

$$\dot{U} = F^{\mathrm{norm}} \times F^{\mathrm{marker}} \times F^{\mathrm{top}}, \tag{5.2}$$

where $\times$ denotes elementwise multiplication. The three terms have the roles of restraining local growth of connections (F^{norm}), keeping similarities of chemical markers on both sides of a link low (F^{marker}), and introducing topological interactions (F^{top}). Thanks to the multiplicative combination, no "tuning" of the relative contributions of the terms is required; the mechanism works for different network sizes without need for adjusting many parameters.

The term

$$F^{\mathrm{norm}}_{i,j} = d - \sum_{\tilde{j}} C_{i,\tilde{j}} \tag{5.3}$$

is a factor that tends to keep the sum of all efferent links from any position i close to a desired value d. Once the combined link strengths exceed d, F^{norm} turns negative, thus letting the respective link shrink. Thus F^{norm} introduces competition among the links extending from any one neuron. Physiologically, this can be easily realized by a global signal in the soma of the neuron that keeps track of the overall strength of all its projections.

The term

$$F^{\mathrm{top}}_{i,j} = \beta(C_{i-1,j-1} + C_{i+1,j+1}) + G_{i,j} \tag{5.4}$$

combines two different topological influences, their relative strength weighted by the parameter β. The first part adds cooperation between parallel neighboring links. This could be realized by cooperative fiber-fiber interactions as discussed in Section 5.1. While this term is not strictly necessary for the mechanism to work, it improves noise robustness of the whole process (see Section 5.3.1). The second term G favors the growth of links to the corresponding position in the next layer ($i-j=0$) over links to faraway positions. We assume it here to be a bounded hyperbolic function of position difference of the two end nodes:

$$G_{i,j} = \frac{\gamma}{|i-j|+\gamma}, \tag{5.5}$$

with γ defining the steepness (see Figure 5.3). G is necessary to tell the ontogenetic mechanism how to align the coordinate systems of the layers it is connecting. A possible way of implementing this term is to first allow development of a point-to-point mapping (e.g. through the mechanism of Willshaw and von der Malsburg, 1979), which then serves as a guidance for the growth of a routing connectivity. This means that the axons have already found their coarse target location, while the *axon collaterals*, the branched terminals of the axon, have not yet finalized their connections.

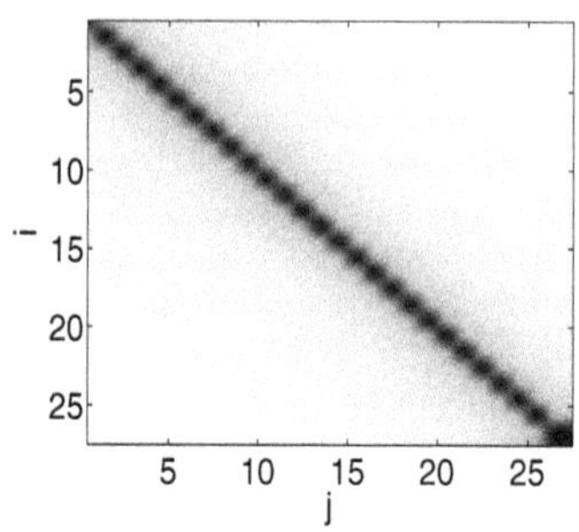

Fig. 5.3 The term G helps to align the coordinate systems of subsequent layers by favoring links between nodes with corresponding positions (middle diagonal) over links between distant nodes. $\gamma = 0.6$ like in the simulations of Section 5.3. Black represents large values of G.

The term F^{marker} represents the "heart" of our ontogenetic mechanism. It makes a link's change sensitive to the similarity of chemical markers in the two nodes it connects. These markers are channeled from the input layer to higher levels by the very connectivity C whose growth in turn they influence. We assume each node of the input layer to contain a different type of chemical marker t_i (for a discussion of the plausibility of this and possible alternatives, refer to Section 5.4). In matrix notation this means that the marker distribution in layer 0 is the identity matrix, $M^0 = I_{N\times N}$, with the marker types on the 1st and the node location on the 2nd dimension. Markers are then transported to higher layers via the existing links C:

$$M^{\kappa+1} = M^{\kappa} C^{\kappa} \ . \tag{5.6}$$

To calculate F^{marker}, we first define a similarity term

$$F_{i,j}^{\text{sim},\kappa} = \sum_{\mu} M_{\mu,i}^{\kappa} (M_{\mu,j}^{\kappa+1} - C_{i,j}^{\kappa} M_{\mu,i}^{\kappa}) \ , \tag{5.7}$$

which is the similarity (dot product) of the marker vector on the presynaptic side with that portion of the marker vector on the postsynaptic side that was *not carried there by the link itself* (μ is an index for the marker type). Therefore, the similarity term signals to the link how well the routes between the part of input space it "sees" and its target node are already being served by other links (see Figure 5.4). The role of F^{marker} is to let a link grow only if its similarity term is not too large. We therefore set

$$F_{i,j}^{\text{marker}} = 1 - H(F_{i,j}^{\text{sim}} - \alpha) \ , \tag{5.8}$$

with $H(\cdot)$ denoting the Heaviside function and a fixed parameter α.

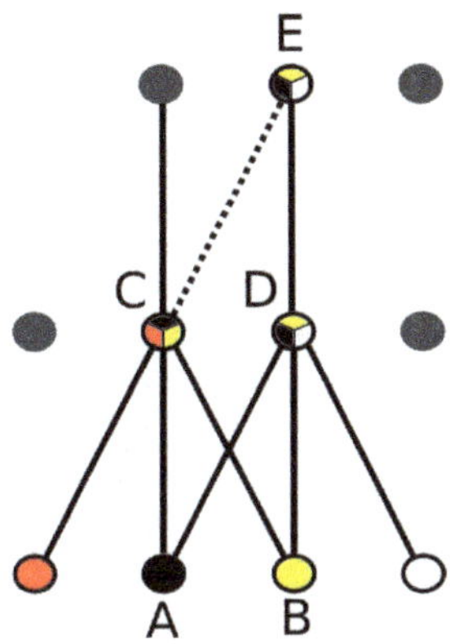

Fig. 5.4 Role of the similarity term. Already well-established links (solid lines) carry markers (represented by different shades of gray) from input nodes A and B to intermediate nodes C and D, and from D to E. Therefore, a weak link C-E (dotted line) finds a marker distribution at its target E that is similar to the one at its origin C. This similarity keeps it from growing. Functionally, this mechanism prevents formation of redundant alternative routes between two points (e.g., A-C-E would be an alternative to A-D-E).

5.3 Results

As example, we choose to investigate the growth of networks containing $k = 3$ link stages. We assume $d = 3$ as target number of links per node (see Eq. 5.3). With 3 link stages, this means that $n = d^k = 27$ input and output nodes can be connected. Equation (5.2) is integrated using the Euler method and the following parameter settings: $s = 30$ (steepness of sigmoid), $\alpha = 0.5$ (threshold for marker similarity), $\beta = 0.6$ (strength of neighbor interaction), $\gamma = 0.6$ (steepness of the hyperbolic term G). A delayed onset of growth at higher stages improves the final results. We choose a delay of 15% and 30% of overall simulation time for the middle and the highest stage, respectively.

When the process starts at the lowermost routing stage, the homeostatic term F^{norm} induces general growth of connections at this first stage. Under the influence of F^{top}, neurons first produce central links to the position in the next layer that corresponds to their own (Fig 5.5a. After this, adjacent links are produced, until F^{norm}

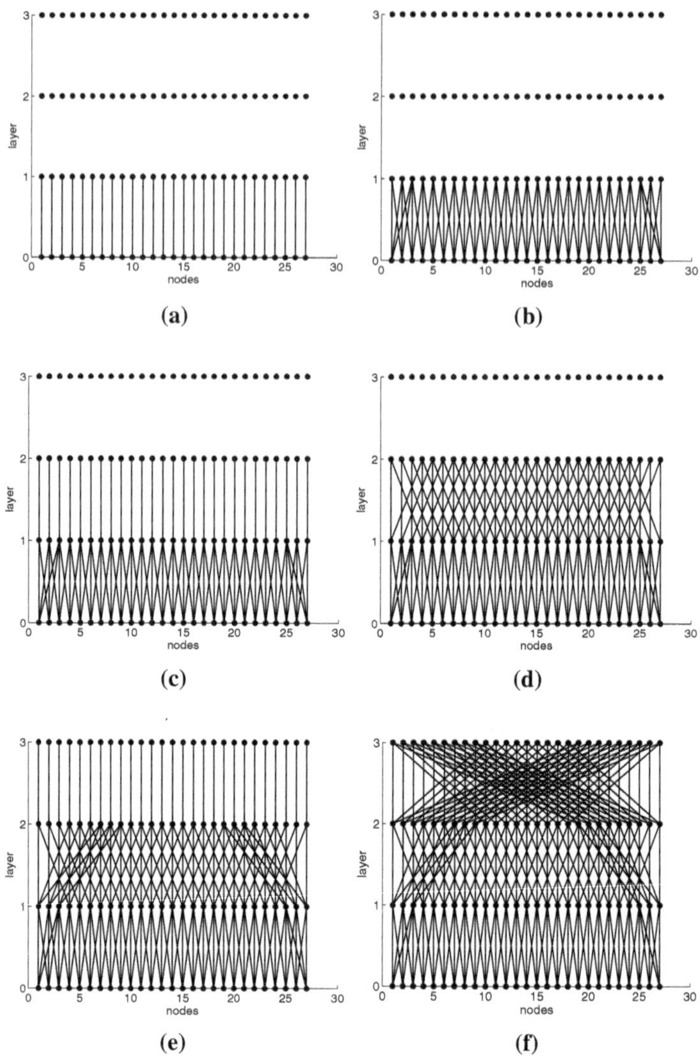

Fig. 5.5 Snapshots of the growth process

stops the growth when the overall strength of each neuron's outward connections has reached its desired value d, here $d = 3$ (see Figure 5.5b). Note that the outermost neurons produce more links slanted to one direction, because they cannot spread their connectivity in both directions. Shortly after, growth sets in at the second stage and produces central links there (Figure 5.5c). The next links here, however, do not grow directly adjacent to the central one. This is because the marker concentrations in the respective nodes are already too similar, so that F^{marker} prevents growth of links here. This situation is visualized in Figure 5.4. The first node in the target layer that does not have any markers in common with the source node is exactly at a distance d from the central location. This is where the next links grow in the second routing stage (Figure 5.5d). Due to the network of links established by the first two routing stages, the chemical markers in the last intermediate layer are spread over a larger distance than in the previous layer. Consequently, once the first central links (Figure 5.5e) have brought markers to the output layers, marker composition in those two layers are similar over larger distances than before. Therefore, the first non-central links at the final stage arise at a distance of d^2 (Figure 5.5e).

The final result of the ontogenetic mechanism is shown in Figure 5.6. Note how the distance between links increases from 1 to 3 to 9 from bottom to top, thus producing non-redundant full connectivity. We can see in Figure 5.6a that the resulting network differs slightly but qualitatively from the manually calculated one of Figure 5.2: there are no wrap-around links (i.e. links from a node on one side of the feature layer to the opposite side of the next layer). Instead, these links appear on the other side of the central link (cf. Figure 5.6b). Interestingly, this new structure produces the same perfect all-to-all connectivity as the one arising from theoretical considerations in Chapter 4, while being biologically more plausible.

The mechanism can also grow routing structures between larger feature layers. For this we only have to adjust the target number of links per node d, without changing any of the other parameters. Figure 5.7 shows simulation results for $d = 5$, i.e. $n = d^3 = 125$ nodes per layer. We see that qualitatively the resulting structure is similar to the one obtained for $d = 3$, except that now each node makes 5 connections to the next layer, with appropriate spacings of 1, 5, and 25 nodes.

5.3.1 *Robustness to Disturbances*

When investigating models for neural development, it is not sufficient to assess their performance under ideal circumstances. Since any developing organism will be subject to certain inhomogeneities of its growth conditions and additional external disturbances, the mechanisms behind the ontogenesis at least of its functionally important structures must be reasonably robust towards such disturbances. Let us examine our model from this perspective.

Figure 5.7 shows the connectivity produced by the model for larger layer sizes and with additional noise in the initial conditions (see below). Clearly, this structure is not as clean as the one in Figure 5.6, with several links not going to the "correct" targets. This means that while some pairs of nodes at the input layer and at the output layer of the Switchyard are connected via two different routes, other such pairs may have no connection. Consequently, the input-output connectivity of the Switchyard, which is defined by the concatenation of all routing stages, $\prod_{\kappa} C^{\kappa}$, is not perfectly homogeneous. For the structure shown in Figure 5.7, the strengths of the input-output connections have mean value $\mu \approx 1$ (which is the optimal value), but a standard deviation of $\sigma \approx 0.15$, which would be zero for a perfect connectivity.

The reason for the uneven final structure lies in the noise that was introduced to the initial link strengths. We chose the initial values of U randomly from the interval $[-16.5 .. -15]$, which means that they contain 10% of additive uniformly distributed noise. Comprehensive investigations with other noise levels show that

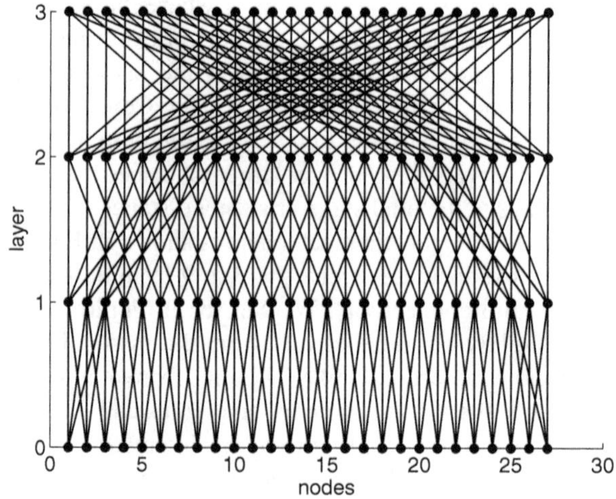

(a)

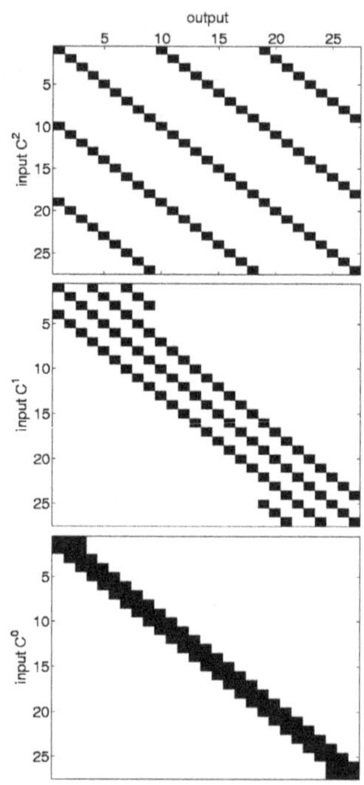

(b)

Fig. 5.6 Results for $n = 27$ nodes per layer and a target number of links $d = 3$. Note that this connectivity does not have "wrap-around" links, i.e. links between one end of the presynaptic layer and the opposite end of the postsynaptic layer. (**a**) Connection structure of the full network. (**b**) Matrices C^{κ} of the full network of (a) shown separately.

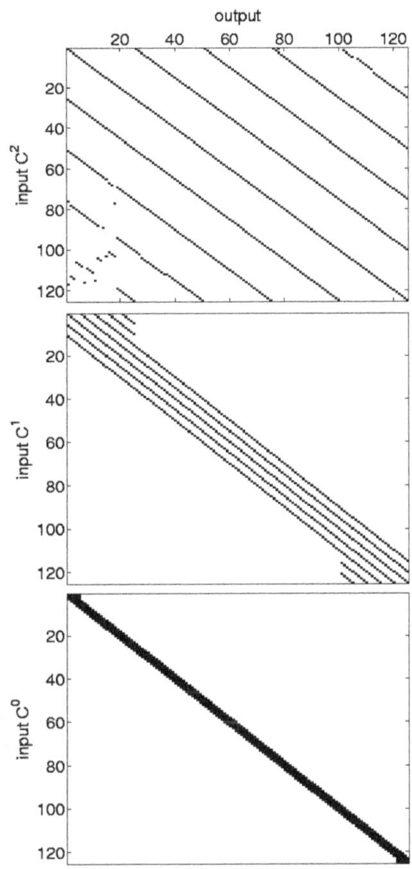

Fig. 5.7 Resulting connection matrices C^{κ} for $n = 125$ nodes per layer and a target number of links $d = 5$. Differently from the situation in Figure 5.6, the initial values of U now contained 10% of additive noise.

the mechanism generally results in a flaw-less connectivity if the initial conditions contain less than $\approx 5\%$ of noise. The growth of smaller networks is far less sensitive to noise. For $n = 27$, up to 20% of additive noise in the initial conditions practically always results in the correct final connectivity (see Figure 5.8).

A reason for this relatively high robustness to noise lies in the topological cooperation between neighboring links (the first component of the term F^{top}). In matrix notation, topological interaction at a stage κ is equal to the connectivity matrix C^{κ} convolved with an oriented kernel G^{top}:

$$F^{\text{top,inter}} = T * C^{\kappa}, \quad T = \frac{1}{2}\begin{bmatrix} 1 & 0 & 0 \\ 0 & 0 & 0 \\ 0 & 0 & 1 \end{bmatrix}. \tag{5.9}$$

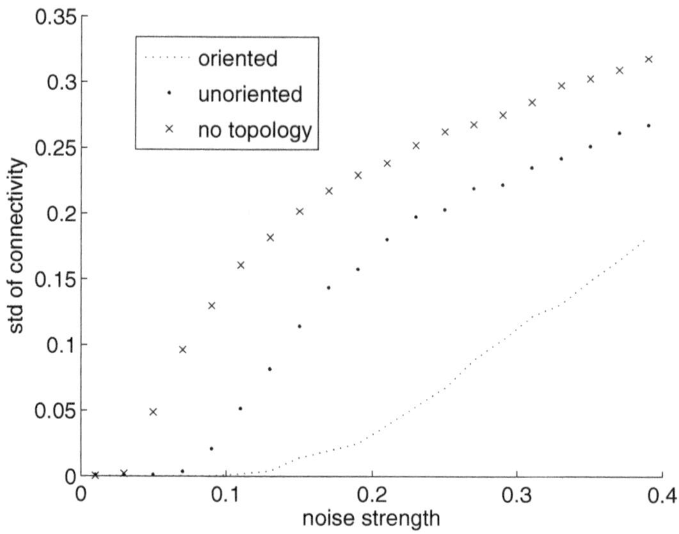

Fig. 5.8 Noise robustness of the ontogenetic mechanism. The figure shows the normalized standard deviation of the resulting input-output connectivity (defined as the concatenation of all routing stages, see text) over relative noise strength in the initial values of U. For a perfect connectivity between input and output of the Switchyard, this std should be zero. Cases shown are oriented topological interaction like F^{top} of the proposed mechanism, unoriented Gaussian interaction, and no topological interaction at all. Values shown are for a Switchyard size of $n = 27$ and have been averaged over 300 runs for each point. Differences between the curves are statistically highly significant.

Previous models (e.g. the model for retino-tectal projections of Häussler and von der Malsburg, 1983) have employed unoriented topologic interaction between links. This was assumed to be biologically more plausible, because it only requires source and target distance, not the orientation of the links (but cf. the fiber-fiber interactions mentioned in Section 5.1 for a possible source of oriented interaction).

Here we assess these two cases (oriented and unoriented interaction) and a third case of no topological cooperation at all in terms of noise robustness. In the third case we replace topological interaction by a scalar constant equal to the mean of topological interaction in the other cases. For unoriented interaction, we use a normalized Gaussian kernel

$$T^{\text{unoriented}} = \begin{bmatrix} 0.0672 & 0.1828 & 0.0672 \\ 0.1828 & 0 & 0.1828 \\ 0.0672 & 0.1828 & 0.06721 \end{bmatrix}$$

with 0 at the central location. As can be seen in Figure 5.8, oriented cooperation is most robust, allowing noise levels of 20% before significant deterioration. Unoriented cooperation produces noisy results from 10% on. When no topological interaction at all is used, final results become noisy already for noise levels of 4%. For every condition (noise level, form of topology), we did 300 runs. A paired t-test on the results shows that the differences are statistically highly significant ($P < 0.001$, but much smaller in most cases). So we can conclude that topological cooperation is not strictly necessary for the principle to work, but it improves noise robustness of the ontogenetic model, making it a far better candidate for explaining the development of connection structures in real organisms.

5.3.2 Growth of Three-Dimensional Networks

We have focused most of our attention on the ontogenesis of flat switchyards between one-dimensional layers. However, the same mechanism can in principle also explain the emergence of routing structures connecting two-dimensional layers of nodes. Figure 5.9 shows a growth process of a Switchyard between layers of $9 \times 9 = 81$ nodes. The parameters used in Figure 5.9 are identical to those of Section 5.2; as the distance between two positions i and j required to calculate $G_{i,j}$ (see Eq. 5.5), we used the Manhattan distance, i.e. the L_1-norm. Unfortunately, these parameters do not produce perfect results in the case of three-dimensional switchyards. As Figure 5.10a shows, not all output nodes are connected to the full input layer in the final network. This changes when the growth process is made fully symmetric by removing boundary conditions in the growth terms. For this, both components of F^{top} have to be adapted. The topological cooperation between neighboring parallel links must be extended to "wrap-around" cooperation between links on opposite ends of a layer that would be neighbors if boundaries of a layer were connected. Likewise, the term G encouraging straight links must not be cut at the boundaries, but must wrap around to the other side of the layer. For such closed boundary conditions, the mechanism produces perfect connectivities also in the three-dimensional case (Figure 5.10b).

One aspect that may be responsible for the less robust growth of three-dimensional networks is the appearance of a rotational symmetry that does not exist for two-dimensional networks. While for one-dimensional layers there are exactly 2 nodes that have a certain distance l from a given node, for two-dimensional layers there is a whole ring of such nodes that sit on the edge of a square centered around that central node. Links to such equidistant nodes receive the same growth signal $G_{i,j}$, which will result in identical growth if everything else is equal. For one-dimensional layers this is no problem, since two links growing to nodes at distance l in opposite directions will not disturb each other. For two-dimensional layers, however, the candidate targets sitting densely on the edge of a square will compete, because a link to one of them would inhibit the growth of others through F^{marker}. Consequently, the decision where links will grow has to happen through spontaneous symmetry breaking in this case. If this symmetry breaking does not happen in a consistent, compatible fashion along the circle of candidate targets, a faulty connectivity will result. Such "symmetric", consistent symmetry breaking is more likely to happen if growth conditions are perfectly symmetric, too, which may explain why wrap-around boundary conditions result in better connectivities.

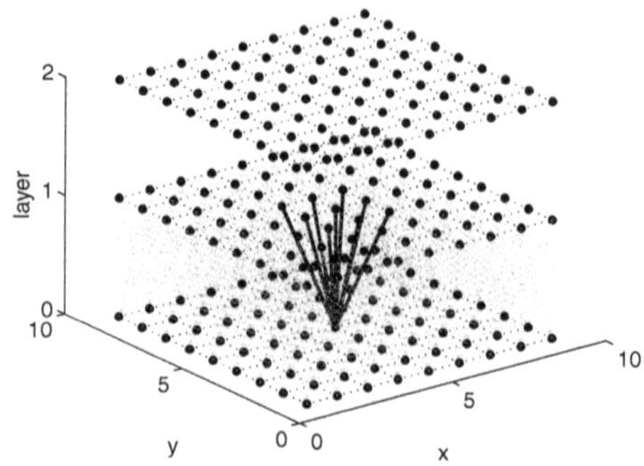

(a)

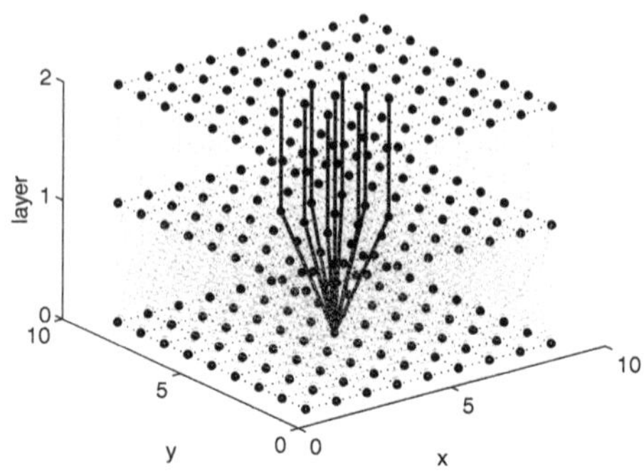

(b)

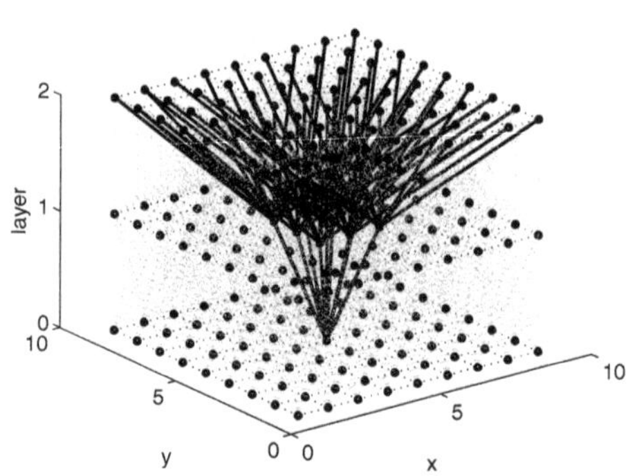

(c)

Fig. 5.9 Growth of three-dimensional networks. Dark lines show the pathways emanating from the central node of the input layer, for three instances during the growth process. All other connections are shifted versions of those (with circular boundary conditions) and are just hinted at by gray dotted lines. Parameters were $n = 81$, $d = 9$.

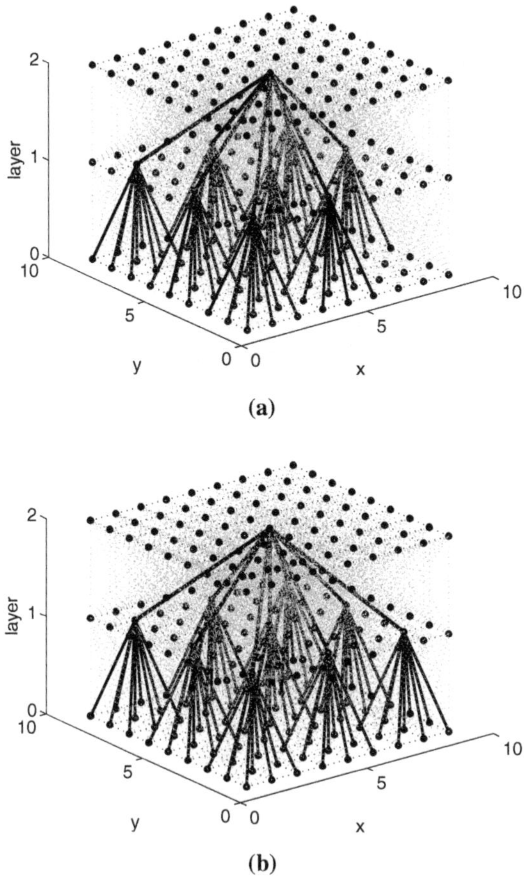

Fig. 5.10 Three-dimensional results without (a) and with wraparound (b) boundary conditions

5.4 Other Potential Mechanisms

One key assumption of the growth model as presented in Section 5.2 is the existence of a unique chemical marker at every input node. At first glance, this may appear like a very unrealistic requirement. Recent findings, however, show that the development of neural connectivity structures does indeed require the existence of thousands of different marker substances. Hattori et al. (2009), e.g., report that Drosophila generates 19,008 different versions of a marker substance that guide the growth of neural connections (in this case, they keep branches of a neuron from connecting back to itself). The authors find that several thousand types of the substance must be expressed in an animal to ensure the proper growth of connectivity. Thus, the assumption of having a unique substance for every input node is not completely unrealistic. Moreover, it is quite plausible that these markers should not only prevent axons from connecting back to their source neuron, but could also be diffused over several connection steps to prevent the development of redundant pathways, as assumed here.

Nevertheless, other mechanisms are conceivable that are functionally similar or equivalent to the model presented in Section 5.2 but do not require this strong assumption. One possibility is a mechanism that operates with only a handful of chemical markers. The concentration of each such marker would be maximal at a certain input node and would fall of monotonously to both sides of that node. Consequently, neighboring nodes already at the input layer would have overlapping marker concentrations to some extent. This means, however, that already at the first routing stage the similarity term between some nodes would be > 0. In order to nevertheless allow tightly bundled links to grow at this first stage, the threshold α has to be raised high enough that this minimal similarity does not affect F^{marker} yet. With this parameter adjustment, a mechanism using only a few chemical markers should be qualitatively similar to the original version.

Instead of trying to find a mechanism with a limited number of markers, however, it is also possible to avoid the use of chemical markers altogether and reformulate the original mechanism of Section 5.2 completely in terms of activity correlation. For this we assume that all input nodes are spontaneously active, $\xi^0_\mu(t)$ representing the activity time course of input node μ. We define a scalar product

$$< \xi, \zeta > = \frac{1}{T} \int_t^{t+T} \xi(\tau)\zeta(\tau) d\tau \tag{5.10}$$

to quantify the similarity between the activities ξ and ζ of two nodes. We require that activities of input nodes are random, uncorrelated with each other, and have constant average energy. Without loss of generality, we can assume this average energy to be normalized such that the scalar product with itself is 1. Under these assumptions, the activities of the input nodes form an orthonormal basis:

$$< \xi^0_\mu, \xi^0_{\mu'} > = \delta(\mu, \mu'). \tag{5.11}$$

Nodes at higher stages are activated by the input layer via their afferent connections. At the first intermediate stage, this activity

$$\xi^1_i = \sum_\mu C^0_{\mu,i} \xi^0_\mu$$

results from the direct links leading from the input stage to node i; for arbitrary nodes of higher stages, it is a general superposition of input activities defined by the connectivity of this node to the input layer:

$$\xi^\kappa_i = \sum_\mu m^\kappa_{\mu,i} \xi^0_\mu, \tag{5.12}$$

$m^{\kappa}_{i,\mu}$ representing the coefficients of this expansion. This expansion is unique due to the ξ^0_t being orthogonal. The similarity of the activities of two nodes in subsequent layers is then given as

$$<\xi^{\kappa}_i, \xi^{\kappa+1}_j> = \sum_{\mu}\sum_{\mu'} m^{\kappa}_{i,\mu}\xi^0_{\mu} m^{\kappa+1}_{j,\mu'}\xi^0_{\mu'}, \tag{5.13}$$

which simplifies to

$$<\xi^{\kappa}_i, \xi^{\kappa+1}_j> = \sum_{\mu} m^{\kappa}_{i,\mu} m^{\kappa+1}_{j,\mu} \tag{5.14}$$

due to the orthonormality of the input layer activities (Eq. 5.11). Comparing (5.14) to the term $\sum_{\mu} M^{\kappa}_{\mu,i} M^{\kappa+1}_{\mu,j}$, which represents chemical marker similarity in (5.7), reveals that the mechanisms for expressing similarity between two nodes are equivalent whether chemical markers or activity correlations are used. Consequently, we can define F^{sim} and then F^{marker} in exactly the same way as done in (5.7) and (5.8), only that now the functional mechanism is neural activity instead of chemical markers. The terms $F^{\text{norm}}_{i,j}$ (defined in Eq. 5.3) and $F^{\text{top}}_{i,j}$ (defined in Eq. 5.4) can remain unchanged since they do not involve chemical markers at all. In this way, the whole growth mechanism has been reformulated in terms of activity correlation without functional changes.

5.5 Conclusion

In this chapter we have developed a neurally plausible ontogenetic mechanism modeling the formation of routing circuits in the brain. The mechanism requires only signals that are available locally at the source and/or target of the respective connection. Interestingly, its results turn out to be biologically more plausible than equivalent connectivities derived from mathematical analysis in Chapter 4 (compare Figs. 5.2 and 5.6a). While the mechanism is useful to understand the development of certain wiring structures of the brain, it may also turn out to have technological applications like the automatic wiring of computer networks.

The functional model presented here may actually be realized by different physiological mechanisms. As presented in Section 5.2, the model requires the existence of a unique chemical marker at every input node. While this assumption may appear unrealistic at first, it is supported by recent findings (Hattori et al., 2009) showing that the stable development of neural connectivity structures does indeed require the availability of thousands of different marker substances. In Section 5.4 we saw, however, that our ontogenetic model can also function with only a handful of chemical markers, or it can operate without any chemical substances, based purely on the correlation of spontaneous firing. It is likely that in reality all of these mechanisms are employed to achieve stable growth, maybe with different importance at different stages of development.

Previous ontogenetic models have mostly focused on retino-tectal projections (e.g., Willshaw and von der Malsburg, 1979; Weber et al., 1997). One of the few exceptions are Linsker's models for the development of receptive fields in V1 (Linsker,

1986). Therefore the study of more complex structures like the Switchyards investigated here promises to yield important new insights, but on the other hand it also has much poorer experimental foundations than the intensively studied retino-tectal projection. A lot of anatomical and physiological work on the ontogenesis of connections in cortex (as opposed to subcortical areas) is necessary to provide a tighter scaffold for future models.

So far, we have focused on the optimal structure (Chapter 4) and growth (this chapter) of Switchyards. In the final chapter, we will investigate how such structures can be used in combination with the ideas of Chapter 3 to make progress towards realistic models of visual object recognition.

References

Hattori, D., Chen, Y., Matthews, B.J., Salwinski, L., Sabatti, C., Grueber, W.B., Zipursky, S.L.: Robust discrimination between self and non-self neurites requires thousands of Dscam1 isoforms. Nature 461, 644–648 (2009)

Häussler, A., von der Malsburg, C.: Development of retinotopic projections: an analytic treatment. J. Theoret. Neurobiol. 2, 47–73 (1983)

Holt, C., Harris, W.: Position, guidance, and mapping in the developing visual system. J. Neurobiol. 24, 1400–1422 (1993)

Linsker, R.: From basic network principles to neural architecture (series). Proceedings of the National Academy of Sciences 83, 7508–1512, 8390–8394, 8779–8783 (1986)

Morton, N.: Parameters of the human genome. Proc. Natl. Acad. Sci. U.S.A. 88, 7474–7476 (1991)

Olson, M., Meyer, R.: The effect of TTX-activity blockade and total darkness on the formation of retinotopy in the goldfish retinotectal projection. J. Comp. Neurol. 303, 412–423 (1991)

Rosoff, W.J., Urbach, J.S., Esrick, M.A., McAllister, R.G., Richards, L.J., Goodhill, G.J.: A new chemotaxis assay shows the extreme sensitivity of axons to molecular gradients. Nature Neuroscience 7, 678–682 (2004)

Schmidt, J., Buzzard, M.: Activity-driven sharpening of the retinotectal projection in goldfish: development under stroboscopic illumination prevents sharpening. J. Neurobiol. 24, 384–399 (1993)

Schmitt, A., Shi, J., Wolf, A., Lu, C., King, L., Zou, Y.: Wnt-Ryk signalling mediates medial-lateral retinotectal topographic mapping. Nature 439, 31–37 (2006)

Sperry, R.W.: Chemoaffinity in the orderly growth of nerve fiber patterns and connections. Proc. Natl. Acad. Sci. U S A 50, 703–710 (1963)

Weber, C., Ritter, H., Cowan, J., Obermayer, K.: Development and regeneration of the retinotectal map in goldfish: A computational study. Phil. Trans. Roy. Soc. Lond. B. 352, 1603–1623 (1997)

Willshaw, D.J., von der Malsburg, C.: A marker induction mechanism for the establishment of ordered neural mappings: its application to the retinotectal problem. Philos. Trans. R Soc. Lond. B Biol. Sci. 287(1021), 203–243 (1979)

Wolfrum, P., von der Malsburg, C.: A marker-based model for the ontogenesis of routing circuits. In: de Sá, J.M., Alexandre, L.A., Duch, W., Mandic, D.P. (eds.) ICANN 2007. LNCS, vol. 4669, pp. 1–8. Springer, Heidelberg (2007)

Chapter 6
Putting the Pieces Together: Recognition with Switchyards

In the previous chapters, we have developed several building blocks for a neural, correspondence-based vision system. In Chapter 3, we discussed how patterns can be compared by finding corresponding points through an all-to-all connectivity, and we showed how these principles can be extended into a neural face recognition system. In Chapters 4 and 5, we investigated how neural patterns can be connected more efficiently by using multistage Switchyards, ignoring, however, the question how these connections may be controlled dynamically. In this chapter, we will combine these ideas. In Section 6.1, we study how correspondences between patterns can be found if the neural representations of these patterns are not directly connected, but only indirectly via a Switchyard of several routing stages. In Section 6.2, we will show how this idea can be extended into a system that can match *and* recognize patterns from a gallery.

Why should it be necessary to match patterns of neural activity via intermediate stages? The short answer to this is (neural) economy. As discussed in detail in Chapter 4, it would be unrealistically expensive in terms of neural circuitry to connect all potential matching partners via direct connections. Therefore it will be a common situation that the brain has to match patterns that are only indirectly connected via intermediate layers. Switchyards (like the one in Figure 6.1a) can provide such a connectivity using a minimal number of links and intermediate feature nodes.

I believe that this is part of a more general computational principle of the brain. Matching two patterns means finding a mapping or transformation from one "input" pattern to another "target" pattern. If the patterns are connected directly (cf. Sect 3.1), the dynamic links between them have to express this transformation in a single step. However, if the patterns are connected by a Switchyard, then each routing stage only represents a partial transformation, all of which are carried out in series to produce the full mapping. In that sense, multiple stages can generate a huge space of transformations through a *combinatorial code* of their simple mappings. This is computationally advantageous to the direct connectivity case, where every single of these possible transformations needs to be represented explicitly.

But while being more parsimonious in terms of required neural circuitry, using multiple stages to connect patterns introduces additional problems when trying to

P. Wolfrum: Info. Routing, Corres. Finding, and Object Recognition in the Brain, SCI 316, pp. 109–122.
springerlink.com © Springer-Verlag Berlin Heidelberg 2010

match these patterns. When patterns are connected directly, dynamic links between them can directly evaluate the similarities of image points and thus establish a global match between the patterns. But when the patterns are connected only indirectly via multiple stages, then the dynamic links of the first routing stage do not have a clear matching partner for the input pattern, and while these early dynamic links have not converged, later routing stages will not receive the properly transformed input pattern to which they can match the target pattern (this situation is sketched in Figure 6.1). However, as we will see in the following sections, matching via multiple stages can work nevertheless under certain condition.

6.1 Matching of Two Patterns

We consider the case of two patterns connected by a Switchyard of connections as introduced in Chapter 4. For illustration, we will use in this and the following section a Switchyard with $k = 3$ routing stages and a layer size $n = 27$, as shown in Figure 6.1a. However, all experiments shown in the following also work for different network sizes and even for three-dimensional Switchyards.In Chapter 4, we have investigated the topology of these networks in detail. On top of the purely static connectivity considered there, however, we now assume that each connection in the Switchyard is governed by a control unit and thus forms a dynamic link which becomes active based on feature similarity of the points it connects and cooperation with its neighboring links (refer back to Section 3.1 for details). Additionally, the

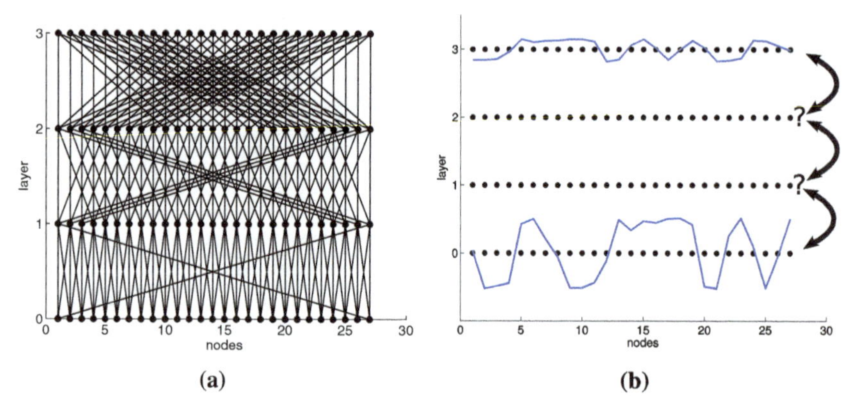

Fig. 6.1 Challenge of matching patterns via a Switchyard. **(a)** The Switchyard used in the experiments in this and the following section has $k = 3$ routing stages and a layer size $n = 27$. Consequently, the fanout at every node is $l = 27^{\frac{1}{3}} = 3$. **(b)** Here, only the feature nodes of the Switchyard are shown, together with the patterns present at input and output of the circuit (the amplitude of the output pattern is lower than that of the input pattern due to different normalizations). The challenge is to match these patterns via a succession of routing stages although a priori not all of the stages have direct access to active patterns that can be used for correspondence finding.

intermediate feature representing layers of the Switchyard now consist each of two separate layers to accommodate separate streams of information flowing upward and downward (sketched in Figure 6.2). Following the nomenclature introduced in Chapter 3 (cf., e.g., Figure 3.9), we call the layers representing bottom-up information *Input Assembly* and the layers containing top-down information *Gallery Assembly*. Like in Chapter 3, both feature and control units follow the modified evolution equation, without competition for the feature units (Eq. 3.2) and with competition in the case of control units (Eq. 3.3). Control units govern both upstream and downstream information flow, and they are activated by evaluating the similarity between

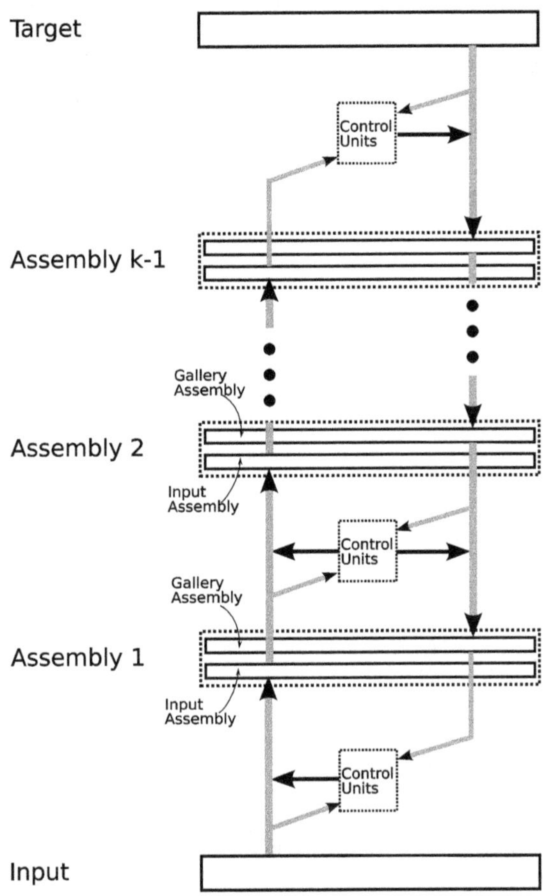

Fig. 6.2 Information flow in the dynamic Switchyard. The input pattern is propagated upstream via the Input Assembly layers, while the target pattern is routed down through the Gallery Assembly layers. All routing is governed by the control units, which in turn are driven by comparison of the activities of the Input Assembly directly below to the Gallery Assembly directly above them.

features in the Input Assembly Layer right below and the Gallery Assembly Layer right above them (again, see Figure 6.2, and also cf. Figure 3.9, where the same entangled principle is used):

$$I_{\text{sim}}(\mathscr{C}_{ij}^{\kappa}) = \mathscr{IA}_j^{\kappa} \mathscr{GA}_i^{\kappa+1}, \tag{6.1}$$

with $\mathscr{C}_{ij}^{\kappa}$ representing the control unit between Input Assembly ($\mathscr{IA}$) unit j in layer κ and Gallery Assembly ($\mathscr{GA}$) unit in layer $\kappa+1$. Additionally, control units of neighboring parallel links cooperate with interaction strength $c_{\text{top},\mathscr{C}} = 1.2$ (see Section 3.4.2 for details on topological interaction among control units).

How can such a Switchyard self-organize into a state that represents a match between the patterns at its input and its output stage? We will focus here on position-invariant matching of one-dimensional visual patterns (like those shown in Figure 6.3) as a proof of principle. The two-dimensional case is a straightforward extension of this, and scale or rotation invariance can be achieved by introducing dedicated dynamic links. In the case of position-invariant matching, the first routing stage of a Switchyard with its tight bundles of connections represents small, but very accurate translations of visual information. The following stages carry out

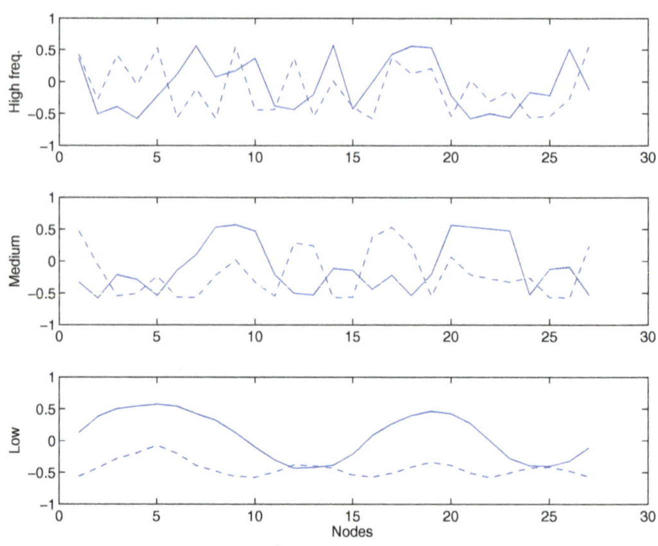

Fig. 6.3 Typical pattern used for matching via a Switchyard. At each position, the pattern consists of 6 components. Every two of them are the real (solid line) and imaginary (dashed line) components of a complex number of unit length. This ensures that the energy of the features is equal at each point. The arguments or angles defining these 3 complex unit vectors (via $z = \exp(i\phi)$) are drawn randomly and independently from a uniform distribution between 0 and 2π for each point and each channel. While the angle pattern of one channel is left unchanged, the others are convoluted with Gaussian filters of width $\sigma = 1$ and $\sigma = 4$ before being converted to complex numbers to produce medium and low frequency signals.

farther reaching but coarser translations, while the highest stage is responsible for global shifts at the scale of the whole image. If these transformations between two patterns are to be estimated more or less independently, both patterns need to contain significant information over a large range of spatial frequencies: low frequencies to match the patterns coarsely at the final routing stage, down to high resolution information to decide on the fine transformations happening at the first routing stage. This condition of having sufficient energy at all frequencies is satisfied for most natural images, but we will discuss the consequences of degenerate cases in Section 6.3.

Nevertheless, we are still confronted with the problem of trying to find this match in a neural fashion via a Switchyard of several stages, each of which can only "see" and match the information directly adjacent to it (see Figure 6.1). However, when all connections are open, the successive stages of a Switchyard act as low pass filters with different cut-off frequencies. Thanks to this property, a Switchyard can actively match visual patterns that have sufficient energy at all frequencies (like those shown in Figure 6.3) as described in the following. Initially, all control units of the Switchyard are active (cf. Chapter 3), which means that all links in the Switchyard can pass information. Since in this state the early stages of the Switchyard act as a low pass, a very blurred version of the input pattern reaches the highest intermediate stage (see Figure 6.4a). Competition between the control units sets in first at the highest routing stages and initiates a correspondence finding process between this low-pass version of the input pattern and the target pattern at the output of the Switchyard (Figure 6.4b). Owing to the significant low frequency components of the patterns, this blurred version of the input is sufficient to find a very coarse estimate of the correct match with the target pattern, just as accurate as it can be represented by

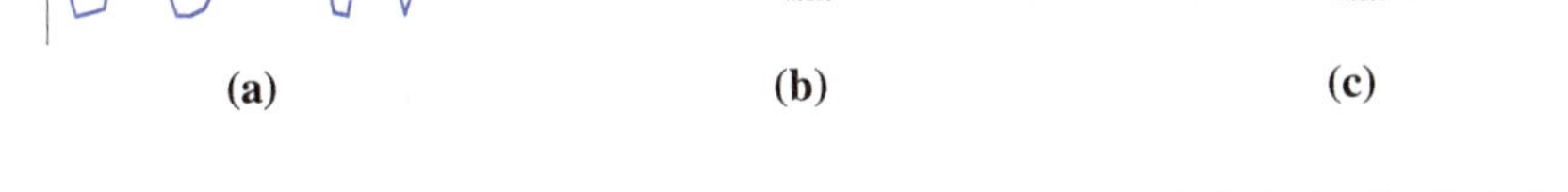

(a) (b) (c)

Fig. 6.4 The principle of matching in a Switchyard. **(a)** Initially, all links in the Switchyard are open. The input pattern is propagated upwards and blurred on the way. **(b)** Matching starts between the target pattern at the output of the Switchyard and the blurred version of the input pattern at the highest intermediate stage. **(c)** Once the links at the highest stage have converged, they send down a properly translated version of the target pattern (9 points to the right in the case shown here) to the next routing stage, where it is matched with a less blurred version of the input pattern to determine the medium-range correspondences. This process travels downstream until the very fine correspondences of the lowest routing stage have been established.

the sparse, far-reaching connectivity provided by the highest routing stage. Once the dynamic links of the highest stage have settled onto this coarse match, they can pass down to the highest Gallery Assembly Layer a version of the target pattern that is already shifted to the approximate position of the input pattern. This pattern can now be matched at a medium routing stage with a less blurred version of the input pattern to establish the medium range translations between the patterns (Figure 6.4c). Finally, a version of the target pattern that is nearly at the same location as the input pattern is passed down to the lowermost intermediate stage, where it is matched with the original input pattern to represent the finest details of the mapping in the activities of the dynamic links of the first routing stage. Snapshots from a simulation of the matching process are shown in Figure 6.5.

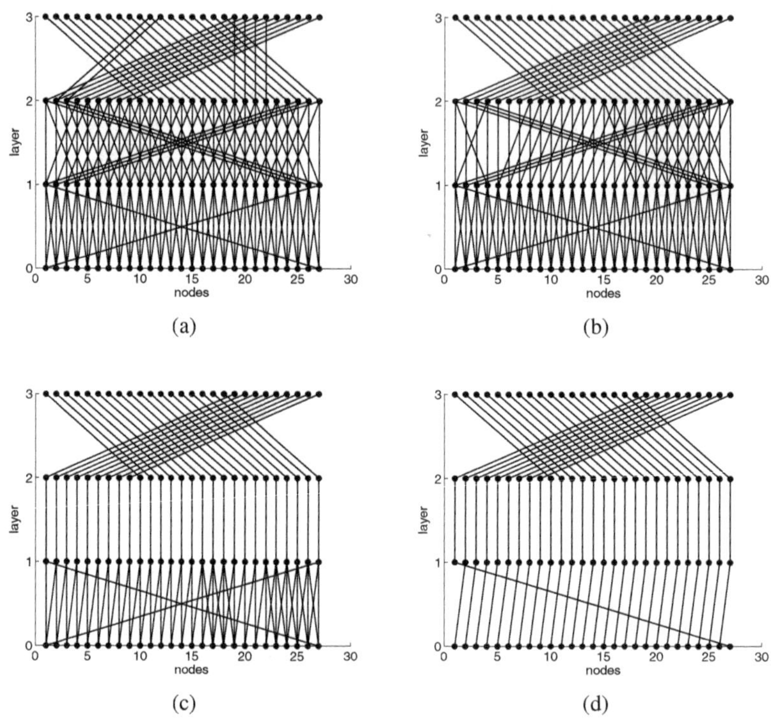

Fig. 6.5 Matching via a Switchyard. The input pattern is shifted by 8 image points to the right with respect to the target pattern (as in Figure 6.1b). Images show the activity of the dynamic links at different stages of the matching process. **(a)** Matching at the highest routing stage has just started. **(b)** The highest routing stage has converged to a shift of 9 image points, while the process has begun at the middle stage. **(c)** The middle stage has produced a straight match, and matching has set in at the lowest stage. **(d)** All dynamic links have converged, representing a transformation that maps the input pattern 8 points to the left.

Comparison to Olshausen's Work

In his Phd thesis, Bruno Olshausen (1994) derived analytically how a routing circuit (the famous *Shifter Circuit*) can autonomously focus on a Gaussian blob. This process is related to the pattern matching via Switchyards described above, and we will see that both processes are functionally very similar. There are some important differences, however, which make the implementation via Switchyards more realistic.

To analytically derive a mechanism for blob focussing, we define an error function E_{blob} that quantifies the mismatch of the output pattern of the routing circuit I^k with a target blob G. If we restrict ourselves to circuits with two routing stages (as Olshausen did), the error function is defined as (the following formulae have been adapted from Chapter 2 of (Olshausen, 1994) to the nomenclature used in this book)

$$E_{\text{blob}} = -\sum_i \mathscr{I}_i^2 G_i, \tag{6.2}$$

with $\mathscr{I}_i^2$ the activity at the ith position of the 2nd (=output) stage of the routing circuit. Note that a Shifter Circuit routes information only in the feedforward direction. We assume here that visual information is propagated linearly and without delay by the routing circuit:

$$\mathscr{I}_i^2 = \sum_j \mathscr{C}_{ij}^1 \mathscr{I}_j^1 = \sum_j \sum_k \mathscr{C}_{ij}^1 \mathscr{C}_{jk}^0 \mathscr{I}_k^0, \tag{6.3}$$

with $\mathscr{C}^\kappa$ representing the control unit activities at stage κ. With this, the derivative of (6.2) with respect to the control units at the second routing stage is

$$\frac{\partial E_{\text{blob}}}{\partial \mathscr{C}_{ij}^1} = -\mathscr{I}_j^1 G_i, \tag{6.4}$$

and the derivative with respect to the lower routing stage results as

$$\frac{\partial E_{\text{blob}}}{\partial \mathscr{C}_{jk}^0} = -\sum_i \mathscr{I}_k^0 \mathscr{C}_{ij}^1 G_i. \tag{6.5}$$

Olshausen then concludes his derivation by defining the blob focussing dynamics through gradient descent on the error function, leading to

$$\tau \dot{\mathscr{C}}_{ij}^1 = \mathscr{I}_j^1 G_i \tag{6.6}$$

and

$$\tau \dot{\mathscr{C}}_{jk}^0 = \sum_i \mathscr{I}_k^0 \mathscr{C}_{ij}^1 G_i. \tag{6.7}$$

In our approach to matching in a routing circuit, on the other hand, there exist separate streams of upward and downward information flow. In steady-state (adiabatic solution of the fast dynamics of (3.2)) the upward stream is described by

$$\mathcal{IA}_i^{\kappa+1} = \sum_j \mathcal{C}_{ij}^{\kappa} \, \mathcal{IA}_j^{\kappa} \tag{6.8}$$

and the downward stream by

$$\mathcal{GA}_j^{\kappa} = \sum_i \mathcal{C}_{ij}^{\kappa} \, \mathcal{GA}_i^{\kappa+1}, \tag{6.9}$$

starting with

$$\mathcal{IA}^0 = I$$

as the input pattern and

$$\mathcal{GA}^k = G$$

as the target pattern (in this case a Gaussian blob). As discussed above, the dynamics of our control units are driven by—besides topological cooperation—the similarity of the Input Assembly location right below and the Gallery Assembly location right above the link (see Eq. 6.1). In the case of a two-layer Switchyard, this yields

$$I_{\text{sim}}(\mathcal{C}_{ij}^1) = \mathcal{IA}_j^1 \, G_i, \tag{6.10}$$

which is identical to the right-hand-side of dynamics (6.6), and

$$I_{\text{sim}}(\mathcal{C}_{jk}^0) = I_k \, \mathcal{GA}_j^1, \tag{6.11}$$

which is equivalent to the rhs of (6.7), since

$$\mathcal{GA}_j^1 = \sum_i \mathcal{C}_{ij}^1 G_i$$

(compare Eq. 6.9).

So in principle, both approaches produce equivalent results when applied to blob focussing, if we simplify (i.e. linearize) the dynamics of the decision units in the Switchyard . aside details of the underlying dynamics. But there is one fundamental difference between both systems that has important implications: Shifter Circuits only route visual information in one direction, from the input to the output stage. The top-down stream that exists in our Switchyards is not explicitly implemented. However, this feedback information is required by the dynamics (6.7), which means that the control units at the lower stages of a Shifter Circuit would need additional, dedicated connections to both the target pattern G and the control units of all stages above them. This provokes the same critique as backpropagation training of multilayer perceptrons: it requires non-local information and therefore is biologically less plausible than a solution using only locally available signals. One possibility to reduce the number of those intricate connections at least a bit would be to do away with the direct connections to the target pattern by hardwiring its values as weights into the connections to the control units of higher stages. This would entail, however, that the circuit now could only use a predefined Gaussian blob as target pattern. While this may allow segmentation in very simplified situations, it will—as already noted by Olshausen (1994)—usually not provide enough details to perform model-based recognition on top of a routing circuit. Moreover, the idea of an explicit downward stream of information agrees with findings that visual imagination evokes activity at "early" visual areas (Kosslyn et al., 1999; Goebel et al., 1998), and it follows our general approach of regarding vision as a generative process (see discussions in Chapter 2 and Section 3.6).

6.2 Recognition from a Gallery of Patterns

Let us now investigate how Switchyards can be incorporated into an invariant object *recognition* system. In principle, this is a straightforward combination of the ideas derived in Section 6.1 and the system developed in Chapter 3. We will demonstrate the essential functionality of the approach with the recognition of one-dimensional patterns from a gallery. Position invariance is achieved by the same 3-stage Switchyard used in Section 6.1. The system performs recognition from a gallery containing 10 "random" patterns (see Figure 6.7). Topological cooperation (refer to Section 3.4.2 for details) among neighboring parallel links in the Switchyard is $c_{\text{top},\mathscr{C}} = 1.2$, cooperation among neighboring Gallery units representing the same pattern has the same value $c_{\text{top},\mathscr{G}} = 1.2$.

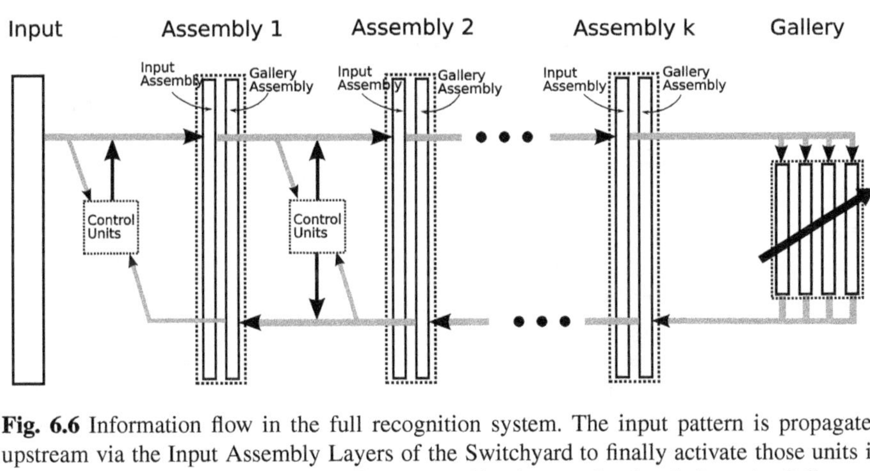

Fig. 6.6 Information flow in the full recognition system. The input pattern is propagated upstream via the Input Assembly Layers of the Switchyard to finally activate those units in the Gallery representing the most similar pattern. Simultaneously, signals from the Gallery are sent to the Gallery Assembly k, where they form a superposition that is routed downstream through the lower Gallery Assembly Layers. All routing is governed by the control units, which in turn are driven by comparison of the activities of the Input Assembly directly below to the Gallery Assembly directly above them. Competition among the control units results in the activation of a unique mapping between Input and Gallery, while competition among Gallery units forces the system to "recognize" one of the patterns stored in the gallery.

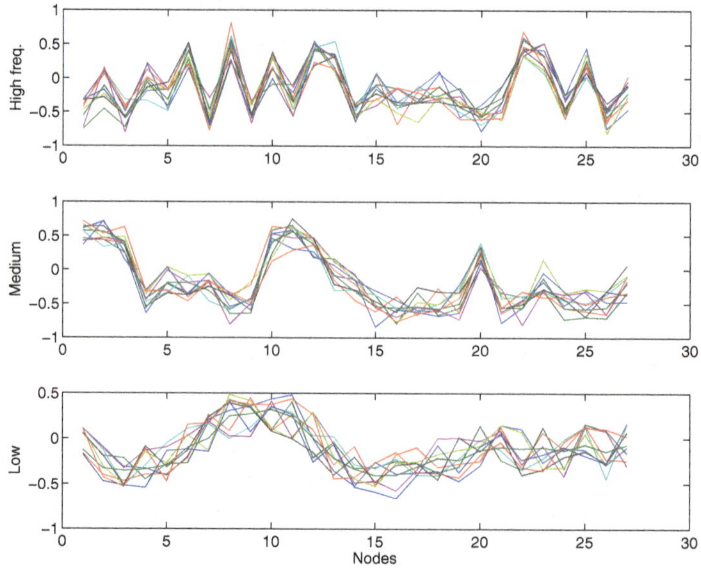

Fig. 6.7 Patterns stored in the gallery. The gallery contains 10 six-dimensional patterns which are defined as described in Figure 6.3. Only the three components corresponding to the real parts of the phase angles are shown here. All patterns are generated from the same template (to emulate the structural similarity of different images coming from the same class of objects), but contain 20% of individual, uncorrelated white noise for all three channels.

We provide an input pattern to the system that is identical to one of the patterns of the gallery, except that it is shifted by several image points. The input pattern enters the system through the Input Layer of the Switchyard (see Figure 6.6) and is propagated upwards through the successive Input Assembly Layers. Since initially all control units have a small, non-zero activity, a very lowpass-filtered version of the input reaches the $(k-1)$th Input Assembly, as discussed in Section 6.1 (cf. also Figure 6.4a). The function of the target pattern of Section 6.1 is now taken over by the Gallery Assembly k, which receives input from all Gallery units. Different than before, however, this target pattern will vary over the recognition process as the activities of the Gallery units change. Initially, all units in the Gallery are equally active, just like the control units of the Switchyard. Therefore the highest Gallery Assembly will initially contain an average of the patterns stored in the Gallery. Since those patterns are created from the same template, this average will contain very similar low-frequency components as the pattern at the Input Assembly $k-1$. This low-frequency information is sufficient to find the rough, long-ranging correspondences that are controlled by this highest routing stage (see Figure 6.8a, also compare Figure 6.4b). Once the control units at this highest stage have converged, the average Gallery pattern can be propagated downwards to Gallery Assembly $k-1$ under the correct long-range translation (see Figure 6.9b). There it is used by the second-highest routing stage to find the medium-range correspondences with the

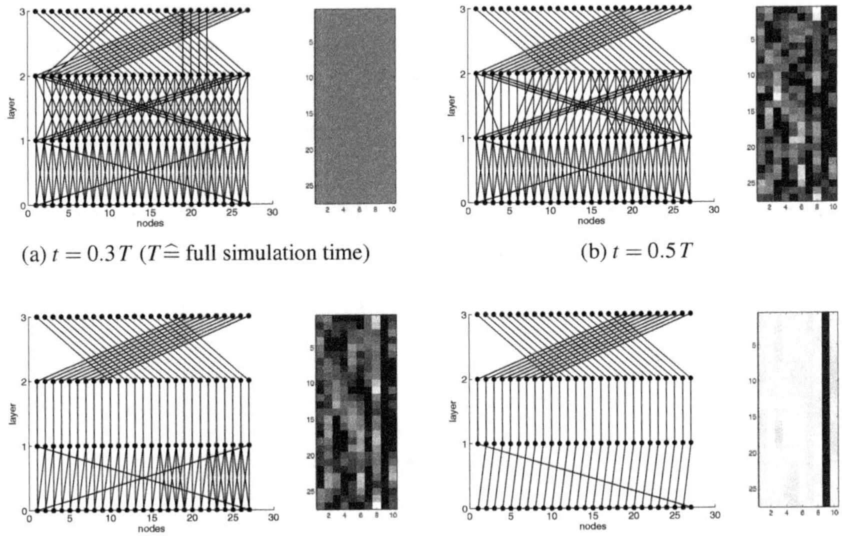

(a) $t = 0.3T$ ($T \widehat{=}$ full simulation time) (b) $t = 0.5T$

(c) $t = 0.7T$ (d) $t = 0.9T$

Fig. 6.8 The process of pattern recognition in the full system. In the example shown here, the system is confronted with an input pattern that is identical with the 9th pattern in the gallery, but shifted by 8 points to the right. The left side of each subfigure shows the active links in the Switchyard connecting the Input with the highest Assembly stage. The right side shows the activation of the Gallery Layer units. Black represents a strong link or strong activity. **(a)** The dynamics of the Gallery start later than those of the highest stage of the Switchyard (with a delay of 40% of the full simulation time T). The highest control units of the Switchyard have already a clear tendency for a shift of 9 image points, while all Gallery units are still in their initial homogeneous state. **(b)** Activity in the gallery has set in, but no trend for any specific pattern is noticeable yet. The highest stage of the Switchyard has mostly converged, while at the medium stage the central links are starting to dominate. **(c)** Matching at the lowest stage is in progress, and a tendency for the ninth gallery pattern is developing. **(d)** The Switchyard has settled to a shift of 8 points, and the Gallery has almost converged to the correct pattern.

version of the input pattern present in Input Assembly $k-2$. The matching process continues downstream through all routing stages until also the very fine correspondences between Input and Gallery Assembly 1 have been established.

In the same way that the decision at the highest routing stage allows the correct downward propagation of the pattern in the highest Gallery Assembly, it also allows the blurred input pattern at Input Assembly $k-1$ to be routed forward under the correct long-range translation to the highest Input Assembly Layer. From there it can serve as point-by-point input to the Gallery, activating more strongly those units that are similar to its own signal. While this similarity is not very crisp initially due to the blurred nature of the signal in the Input Assembly stream, it improves with every routing stage below that finds its correspondences and sends a sharper version of the input pattern upwards. The sharper the pattern in the highest Input

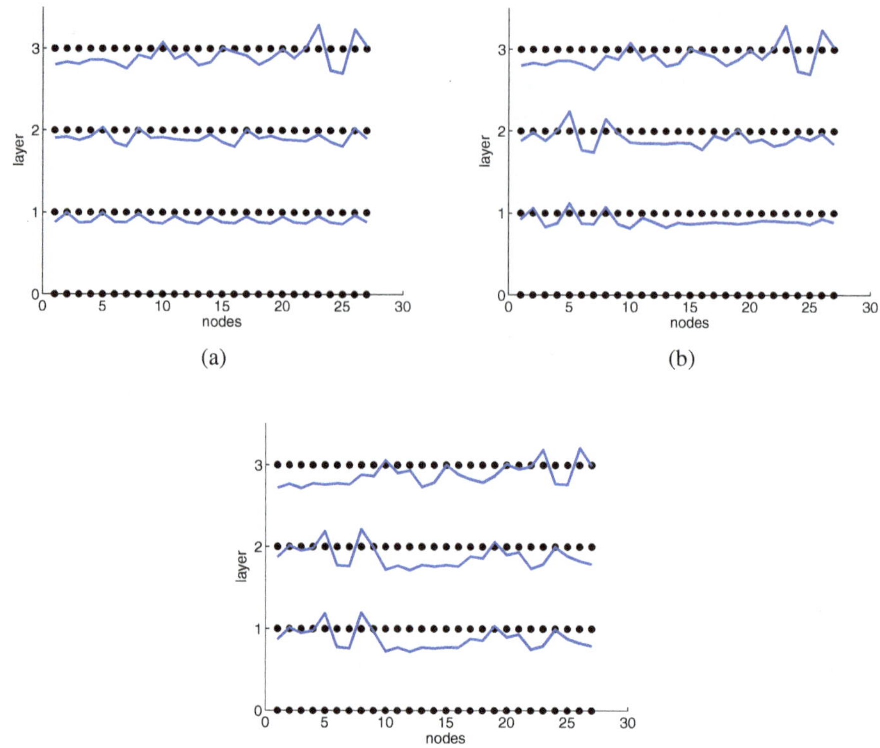

(a) (b) (c)

Fig. 6.9 Signals generated by the Gallery and propagated down the Gallery Assembly stream. **(a)** Initially, the uppermost Gallery Assembly Layer contains an average of all Gallery patterns, while the lower layers only receive lowpass versions of this signal. **(b)** After the links of the highest routing stage have converged, they route a sharp version of the Gallery average to the middle Gallery Assembly. **(c)** In this image, both the routing process and the Gallery competition have mostly finished. The uppermost layer now contains a copy of the chosen Gallery pattern instead of the average of all of them (note the subtle differences, e.g. at nodes 4–7). This pattern is precisely routed to the lower stages.

Assembly, the clearer becomes the decision among the candidate patterns of the Gallery. This in turn changes the pattern sent down through the Gallery Assembly stream from an average of all Gallery patterns towards that pattern which is dominating the competition (see Figure 6.9c). This enhancement of the downstream target pattern is necessary especially for fixing the very fine correspondences at the lowest routing stage. If the recognition process is successful, the whole system ends up in a state where the control units of the Switchyard represent the correct shift from Input to Gallery, while the one Gallery pattern remaining active is the one the system has "recognized" as being most similar to the input pattern (Figure 6.8d). Neighbor-neighbor cooperation between parallel links as well as between Gallery units representing the same pattern helps the system to perform robustly on this challenging "hen and egg" problem even in the presence of noise. Note in Figure 6.7 and by comparing Figures 6.9b) and 6.9c) how small the differences between the Gallery patterns are. This is exactly what "recognition" means: Identifying from a large set of structurally identical images the one that most resembles an input image in all its minute details.

6.3 Conclusion

We have seen in this chapter how the ideas of the previous chapters can be combined into a position-invariant recognition system. We first investigated how the single stage correspondence finding of Section 3.1 can be extended to matching patterns via the intermediate stages of a Switchyard as introduced in Chapter 4. We then proceeded to combine these results with the object recognition ideas of Section 3.4. The resulting system combines the advantages of correspondence-based object recognition (see Chapter 2) with the biologically (and computationally) more plausible idea of achieving full connectivity via a Switchyard of routing stages.

This claim that a system using several layers for correspondence finding is biologically more realistic is backed—additionally to the physiological evidence discussed in Section 4.2—by the fact that the approach breaks down for visual patterns that pose problems for human observers, too. For example, it is quite difficult for humans to recognize *random dot stereograms* (RDSs). RDSs are a superposition of two fine-grained patterns (see Figure 6.10) the difference of which encodes a virtual depth profile of the image. When merged properly, they give rise to a very strong sensation of depth. Computationally, recognition of RDSs is equivalent to merging two normal stereoscopic images, and both these tasks can be formulated as correspondence finding problems between two patterns as discussed throughout this book (the directions of the active links between the patterns will encode the depth profile of the object). Consequently, the single stage matching process of Section 3.1 would perform equally well and fast for RDSs as for other visual patterns. The difference between RDSs and "normal" visual information is, however, that the former do not contain any low-frequency cues that can be used to establish coarse correspondences between the two patterns. Therefore, the multistage matching process introduced in Section 6.1 would not be able to successfully match the information contained in a random dot stereogram, since it would lack the low-frequency correlations that are necessary to initiate matching at the higher routing stages in the first place. Likewise, many humans cannot recognize these random dot stereograms at all, and if we can, it is a slow process that requires patiently staring at the image and waiting for the brain to make sense of the correlations among the random dots. This example illustrates the limits of multistage routing, which may be quite similar to the limits of human visual perception.

Fig. 6.10 Random dot stereogram. When looking at this figure and trying to focus behind it by lightly squinting outwards, one may perceive a three-dimensional shape (in this image a relief of the Mandelbrot set).

References

Goebel, R., Khorram-Sefat, D., Muckli, L., Hacker, H., Singer, W.: The constructive nature of vision: direct evidence from functional magnetic resonance imaging studies of apparent motion and motion imagery. Eur. J. Neurosci. 10, 1563–1573 (1998)

Kosslyn, S.M., Pascual-Leone, A., Felician, O., Camposano, S., Keenan, J.P., Thompson, W.L., Ganis, G., Sukel, K.E., Alpert, N.M.: The role of area 17 in visual imagery: convergent evidence from PET and rTMS. Science 284, 167–170 (1999)

Olshausen, B.: Neural Routing Circuits for Forming Invariant Representations of Visual Objects, PhD thesis, California Institute of Technology (1994)

Chapter 7
Discussion and Outlook

In the first two chapters of this book, we claimed that the matching of spatially distributed patterns is central to the function of our brain, and we named two requirements for this: suitable connection structures between the units representing those patterns, and a basic mechanism—plus its biological realization—for matching patterns.

We presented this basic mechanism, dynamic link matching, in Section 3.1 and showed in the following sections how it can be realized in a neurally plausible way and how it can be extended into a functioning, competitive face recognition system. Regarding connection structures, in Chapter 4 we derived a multistage architecture called Switchyard that provides (indirect) full connectivity between two areas using a minimal number of neural resources. In Chapter 5, we investigated the growth of connections in the brain and found an ontogenetic mechanism that produces in a self-organized, autonomous way connection structures that are nearly identical and functionally equivalent to those derived analytically in Chapter 4. Finally, in Chapter 6 we showed that object recognition can actually work by matching patterns across the multiple layers of a Switchyard.

But why should the matching of patterns via intermediate stages be a general principle of brain function, as we claimed at the beginning of the book? Additionally to the arguments discussed in Chapter 2, the findings presented in the subsequent chapters support this conviction in several ways.

First of all, an object recognition system built on these principles is inherently *generative*. It requires that information flows not only upstream from the input, but also downstream from an internal memory (see Figure 6.6), thus generating at "lower" stages explicit representations of its percepts and decisions in higher areas (see e.g. the patterns propagated down the Gallery Assembly stream in Figure 6.9, or the content of the Gallery Assembly in Figure 3.13 changing from an average to the recognized face). This entails that the system of Chapter 3 cannot only detect general faces, but can in a very natural way be biased e.g. for "female" faces or search for a specific individual. The system can exhibit attentional effects through priming of its routing units on the input side, as well as expectation through priming of its memory representations. It can also replicate effects like "day dreaming"

P. Wolfrum: Info. Routing, Corres. Finding, and Object Recognition in the Brain, SCI 316, pp. 123–126.
springerlink.com © Springer-Verlag Berlin Heidelberg 2010

and shows activity at lower stages generated from memories, as has been observed in physiological experiments. In that sense, feedforward and feedback streams, external input and internal memory become two equally important sides of the same coin.

Another advantage is the *combinatorial coding of transformations* introduced by having multiple routing stages. When matching two patterns, the circuitry connecting them must be able to carry out a multitude of transformations to map the patterns onto each other in spite of their different extrinsic properties (position, scale, color, etc.). If the patterns are connected directly, then dynamic links between them must be able to express these transformations in a single step. However, if the patterns are connected by a Switchyard, then each routing stage only represents a partial transformation, all of which are carried out in series to produce the full mapping. In that sense, multiple stages can generate a huge space of transformations through a combinatorial code of their simple mappings. This is computationally more economic than the direct connectivity case, where every single of these possible transformations needs to be represented explicitly. We saw in Chapter 4 that Switchyards, which are capable of performing arbitrary translations as well as many types of scaling and rotation, are much cheaper in terms of required neural circuitry than if these transformations were to be carried out in a single step (cf. Figure 4.2). For realistic vision scenarios with even more types of transformations, it would be imperative to carry them out in a circuit of sequential stages. Of course, the division of the matching process into several stages has its drawbacks. To allow it to function, the different stages have to be able to operate at least partially independently. If the initial loss of fine-scale spatial information makes the visual input insufficient for a successful matching at higher stages, then the multistage approach will not work, as we saw in the case of the random dot stereogram in Section 6.3. For such problems, the brain probably resorts to serial processing of small patches of the input image at a time, thus effectively trying out all combinations of control units at the different stages and hence negating the advantages of multistage routing. Perception becomes slower and requires conscious effort.

Finally, vision systems based on matching via several stages could offer a *synthesis* of the feature-based and correspondence-based "philosophies" in vision: Initially, input information flows upstream through the Input Assembly stages with all routing gates open (see Figure 6.4a). This first bottom-up sweep is very similar to the behavior of feature-based systems. If the free flow of information were replaced by a maximum pooling at each receiving feature node, such a system would initially respond similarly to models like those of Riesenhuber and Poggio (1999). The matching process that follows this first sweep, on the other hand, is inherently correspondence-based. When matching has finished on all routing stages, the final situation is given by a complete match across all stages that, different to feature-based approaches, does not lose information about the properties of the input pattern. In that sense, systems built on the principles of Chapter 6 could reconcile the findings of extremely fast, unconscious recognition (Thorpe et al., 1996) with the advantages of correspondence-based or generative systems (very exact recognition, reasoning about a percept, etc.).

Of course, the results presented in this book can only be a starting point in this direction. Especially the unification in Chapter 6 of the principles derived in the preceding chapters is only demonstrated on toy examples. Apart from dealing with more realistic data here, the system needs to be extended towards being able to recognize images from more than a single category. For object categories with a clear topological layout, this should in principle be possible as sketched in Figure 7.1. The system would first coarsely match the input information at the highest Assembly stage with averages generated from the respective category memories to switch on the dynamic links to only one category ("flowers" in the case shown in Figure 7.1). This would result in an average pattern in the top-down Gallery Assembly stream generated only from the chosen category. From then on, matching would proceed as described in Chapter 6 to determine the individual object within that category.

An unrealistic aspect of the model so far is the flat structure of the gallery domain. As has been convincingly shown (Biederman, 1987), many objects are recognized as ordered arrays of simpler sub-shapes. Thus, just like we extended the correspondence-finding part of our system from one-step matching to multi-layered Switchyards, a hierarchy should also be introduced on the Gallery side. This would help the system to deal better with images that are not as strictly spatially defined as faces (e.g. objects made up of sub-components, or landscape scenes). Obviously, hierarchical structuring of a memory domain is more challenging than structuring of the correspondence-finding process, where we could exploit the strictly geometric nature of early visual representations.

So there are several different directions that need to be pursued to obtain a fully functional explanation object recognition in the brain. We hope that the building blocks provided by this book—a biologically plausible approach to recognition by finding correspondences between layers of units and an approach to break pattern matching into a Switchyard of several stages—will help on this way.

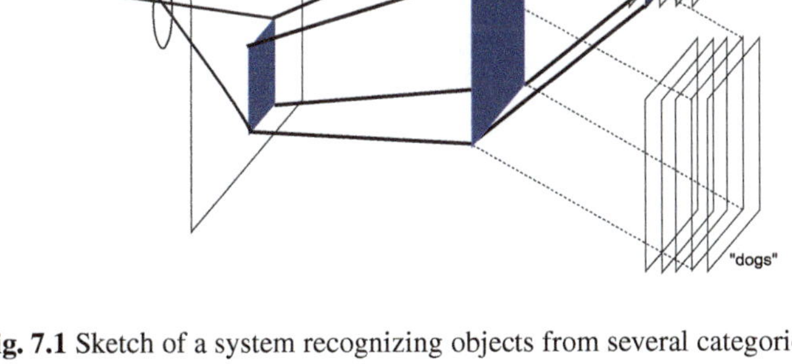

Fig. 7.1 Sketch of a system recognizing objects from several categories

References

Biederman, I.: Recognition-by-components: a theory of human image understanding. Psychol. Rev. 94(2), 115–147 (1987)

Riesenhuber, M., Poggio, T.: Hierarchical models of object recognition in cortex. Nature Neuroscience 2(11), 1019–1025 (1999)

Biederman, I.: Recognition-by-components: a theory of human image understanding. Psychol. Rev. 94(2), 115–147 (1987)

Appendix

A.1 Self-normalization Properties of Columnar Dynamics

The dynamics of the column model introduced in Section 3.3 possess automatic activity normalization, as shown in the following. Since the ratio $\frac{T}{\tau}$ is very large, i.e. the time constant is much shorter than the overall simulation time, the unit activities are close to the adiabatic state, i.e. $\frac{d}{dt}x_i \approx 0$.

For the case of $\nu = 0$ (linear representation), the unit dynamics are given by (3.2):

$$\tau \frac{d}{dt}x_i = I_i - x_i \sum_{j=1}^{K} I_j x_j.$$

The steady state of this is

$$x_i = \frac{I_i}{\sum_j I_j x_j}. \tag{A.1}$$

Multiplying both sides with x_i yields

$$x_i^2 = \frac{I_i x_i}{\sum_j I_j x_j},$$

and the sum of this term over all i is

$$\sum_i x_i^2 = 1, \tag{A.2}$$

i.e. for $\nu = 0$ the column activity is normalized to a 2-norm or Euclidean norm of 1. From squaring (A.1) and summing over all units, we get

$$\sum_i x_i^2 = \frac{\sum_i I_i^2}{(\sum_j I_j x_j)^2},$$

and plugging in the normalization property (A.2) yields

$$\left(\sum_j I_j x_j\right)^2 = \sum_j I_j^2.$$

Using this result in (A.1) gives

$$x_i = \frac{I_i}{\sqrt{\sum_j I_j^2}}. \tag{A.3}$$

This means that the steady state of a feature unit is independent of the other unit activities in the column and is the ratio of the unit's input and the energy in the overall column input.

In the case of $\nu = 1$ (WTA behavior), the unit dynamics are

$$\tau \frac{d}{dt} x_i = x_i I_i - x_i \sum_{j=1}^{K} I_j x_j,$$

so summing over all units gives

$$\tau \sum_i \frac{d}{dt} x_i = \sum_i x_i I_i - \sum_i x_i \sum_j x_j I_j = \left(1 - \sum_i x_i\right) \sum_i x_i I_i \stackrel{!}{=} 0 \tag{A.4}$$

in steady state. If there is any activity in the column, i.e. not all unit activities are zero simultaneously, this requires

$$\sum_i x_i = 1, \tag{A.5}$$

which means that for $\nu = 1$ the column activity self-normalizes to a 1-norm (also Manhattan norm in this case, since unit activities are strictly positive) of 1. Consequently, the interaction term $\sum_j I_j x_j$ is the average activity-weighted input to the column. This means that only those unit activities grow whose input is higher than this weighted mean input to the column, otherwise they decay. This lets the weighted input average grow, because the bias shifts towards strong inputs. Eventually, all unit activities decrease to 0 except for the unit with the strongest input, whose activity approaches 1. For the final steady state we can show this by setting the time derivative for a single unit to 0:

$$x_i \left(I_i - \sum_{j=1}^{K} I_j x_j\right) = 0. \tag{A.6}$$

Here we see that for any i, we either have $x_i = 0$, or $I_i = \sum_j I_j x_j$, which can only be true for $x_i = 1$ and *all* other $x_j = 0$ (except for the degenerate case of two or more of the I_i being exactly identical, which would presumably be solved in the brain by spontaneous symmetry breaking). So in the final state, one unit must have activity 1 with all other unit activities at 0.

A.2 Gabor Transform

The Gabor transform dates back to 1946, when it was first proposed by the Hungarian engineer Dénes Gábor. Coarsely speaking, it acts like a localized Fourier transform, representing the frequency content of a window around a certain location of an image. The most common window function—which we use in this book as well—is a Gaussian. An intersting property of the Gabor transform is that among all wavelet transforms, it optimally preserves both location and frequency, making it the best choice for representing local features of, e.g., an image. Gabor wavelets are linear filters and are regarded in computational neuroscience as a good compromise between accuracy and simplicity for describing the responses of simple and complex cells in visual cortex.

The model described in Chapter 3 represents an input image as a grid of Gabor features. At each grid point, it uses a set of Gabor wavelets that appropriately sample orientation (over 8 orientations) and spatial frequency (over 5 scales) space. If V is an image with $V(\mathbf{z})$ denoting the gray-value of a pixel at the geometric position $\mathbf{z} = \binom{x}{y}$, the filter responses $R_i(\mathbf{z})$ are given by:

$$R_i(\mathbf{z}) = \int V(\mathbf{z}')\,\psi_i(\mathbf{z}-\mathbf{z}')d^2\mathbf{z}', \tag{A.7}$$

$$\psi_i(\zeta) = \frac{k_i^2}{\sigma^2}\exp\left(-\frac{k_i^2\zeta^2}{2\sigma^2}\right)\left[\exp(i\mathbf{k}_i\zeta) - \exp\left(-\frac{\sigma^2}{2}\right)\right], \tag{A.8}$$

where we choose

$$\sigma = 2\pi \tag{A.9}$$

to approximate the shape of receptive fields found in primary visual cortex, and where the wave vector is parameterized as

$$\mathbf{k}_i = \begin{pmatrix} k_{ix} \\ k_{iy} \end{pmatrix} = \begin{pmatrix} k_\rho \cos\varphi_\mu \\ k_\rho \sin\varphi_\mu \end{pmatrix}, \; k_\rho = 2^{\left(-\frac{\rho+2}{2}\right)}\pi, \; \varphi_\mu = \mu\frac{\pi}{8}, \tag{A.10}$$

with orientation parameter $\mu = 1,..,8$ and scale parameter $\rho = 1,..,5$. That is, $(R_1(\mathbf{z}),\ldots,R_{40}(\mathbf{z}))$ is a vector of Gabor filter responses in which each entry corresponds to one of the 40 combinations of ρ and μ. As feature values we use the magnitude

$$\mathscr{J}_i^p = |R_i(\mathbf{z}_p)|, \tag{A.11}$$

thus ignoring Gabor phase, to model complex cell responses.

Index

G

H

I

L

M

N

O

P

R

S